Contents

Introduction

The Mediterranean region has long been a favoured destination for European bird-watchers, and with good reason. Apart from its enviable climate – to those of us from more northerly latitudes at least – it boasts a stunningly diverse and rich birdlife. For the casual observer, the Mediterranean offers exotic-looking and colourful species aplenty; given the right timing and location, the spectacle of numbers that is migration can be witnessed too. For the more dedicated birder, the region also has its fair share of challenges: there are species so secretive that they almost defy observation, and there are identification challenges over which to puzzle for hours on end. I have been visiting the Mediterranean for the last 30 years and wherever I go I never cease to be amazed at the variety and richness of birdlife I encounter. I hope that some of my enthusiasm for this stunning region is conveyed by this book.

How to Use this Book

This book has been arranged so that the photographs and text for each species are on facing pages; number and letter codes indicate clearly the relationship between the two. The adult plumage is shown for each species unless indicated otherwise. The following abbreviations have been used on the photographs: imm = immature; juv = juvenile; ad = adult; 1st-w = 1st-winter; 2nd sum = 2nd summer; non-br = non-breeding; ad br = adult breeding; subad = subadult; ad w = adult winter.

Each description begins with the bird in question's common English name, followed by its scientific name. There then follows some indication of size – always a useful factor to consider when attempting identification. In most cases the length of the bird is given, where this is the most useful measurement for comparison purposes; the length is measured from the bill tip to the tip of the centre of the tail. However, for species such as birds of prey that are seen mostly in flight, the given measurement is wingspan (WS), this being the distance from wingtip to wingtip.

In order to cram as much information as possible into the text, the species descriptions are written in an abbreviated style. Where sex and age differences exist for a given species, typically the description will describe adult male and female plumages before tackling immature stages. For adults, the most typical plumage for the region is described first, and generally this means breeding or summer plumage. Information is also provided about species' habits, preferred habitats, status within the region and geographical range. Where relevant to identification, songs and calls are also described briefly.

Species Selection

Bordered to the north by Europe and to the south by Africa, it is no wonder that the Mediterranean region as a whole has a rich birdlife. Further contributions to this richness are made by the influences – on both climate as well as species – of the Atlantic Ocean to the west and the land mass of Asia to the east. In addition to its resident birds, the Mediterranean also hosts separate and discrete populations of winter and summer visitors. Furthermore, it lies on the route for vast numbers of migrant birds that breed further north in Europe but that winter to the south in Africa or to the southeast in Asia. This book covers 440 bird species, the list comprising residents in the Mediterranean region as well as regular passage migrants and migrant summer or winter visitors.

The Region Covered by this Book

As you might guess, the near-landlocked Mediterranean Sea is the centrepiece of the geographical range covered by this book. The coverage extends north to include the Iberian Peninsula in the west and the Black Sea in the east; North Africa forms the southern boundary to the map. In terms of longitude, the Atlantic shores of Iberia and North Africa mark the western extent of the range of this book, while the Middle East defines the east.

A consequence of defining the range covered by this book has been to include areas in addition to those that experience a Mediterranean climate in the strict sense. And so, as a result, true desert birds are included as well as species that are found on mountain slopes for at least part of the year.

Key to Maps

 The maps should be viewed as an indication of range and not an absolute guide. Bird ranges change: vagrancy and dispersal occur and ranges expand and contract over the years.

GREEN – the range in which a species is present year-round.

YELLOW – the range in which a species is present as a summer visitor and in which passage migration can also be observed.

BLUE – the range in which a species is present as a winter visitor and in which passage migration can also be observed.

PINK – the range in which a species appears seasonally as a passage migrant in spring and autumn; it is also the area in which some species breed in small numbers on a more-or-less regular basis; in the case of essentially resident species that are nomadic within the region outside the breeding season, it represents the typical extent of their dispersive movements.

All the birds in this composite photograph are Black Kites. It is sobering to realise that the appearance of individuals of the same species can vary so much depending on your viewing position.

What's in a Name?

Each species entry starts with the most widely accepted modern English name for that bird. Some of the English names used may differ slightly from those that you are familiar with from earlier field guides. For example, in the case of birds such as the Wheatear *Oenanthe oenanthe*, previously known by this name alone, the modern convention of adding the prefix 'Northern' has been adopted. This sensible addition avoids ambiguity and confusion with other wheatear species. You will also find that a few bird names used in this book simply do not exist in works produced as recently as 10 years ago. This is because recent research, often assisted by DNA analysis, has identified areas where species or subspecies status has been ascribed incorrectly in the past. Thus, Southern Grey Shrike *Lanius meridionalis*, once regarded as a subspecies of Great Grey Shrike *Lanius excubitor*, has been elevated in this book to species status and embraces many of the subspecies that formerly fell under the umbrella of Great Grey Shrike. Some of these changes are still the subject of discussion among taxonomic experts and are not accepted by all concerned.

The English name is followed by the bird's scientific name, written in a form of Latin. Not only is this name unique to each species, but it is regarded as something of a constant in the scientific world; changes or modifications occur only when species or genus splits occur, and not for more esoteric reasons. The benefit of having a scientific name for each species is that it is universal and remains the same in any language. In most cases the scientific names used in this book comprise two words: the genus name is given first, followed by the specific name. Closely related species often share the same genus name but, when combined with the specific, their name becomes unique. In the case of a few birds, you will find a third scientific name; this refers to a particular subspecies where this information is pertinent to the region. A subspecies is a visibly distinct population (in terms of plumage) within a species, typically with a well-defined geographical range and separate from the rest of that species.

On the whole, the order in which the birds appear in this book follows the convention adopted by other modern field guides. Thus, the species entries begin with divers and grebes and end with passerines. However, in order to make best use of the photographs, design constraints have dictated the need to make minor changes to the strict order.

Previous given names for the White-throated Kingfisher included White-breasted Kingfisher and Smyrna Kingfisher. However, the scientific name, Halcyon smyrnensis, *has remained unchanged.*

6

The Use of Photographs in Field Guides

Opinions differ as to whether field guides are best illustrated using artwork or photographs. In their favour, photographs are perceived as accurate, capturing a moment in time. Set against them, a major criticism is that not all photographs are ideally suited to field identification. Differences in lighting or pose of similar species can make it difficult, if not impossible, for direct comparisons to be made. Furthermore, financial and time constraints imposed on the design of photographic field guides in the past have often dictated the inclusion of only one image per species. In the case of passerines or ducks, for example, typically this would have meant using a male in breeding plumage and ignoring – in terms of illustrations – female and non-breeding plumages altogether. With birds whose appearance is determined by a complex regime of seasonal and age differences, clearly this is less than helpful in the field.

However, over the last five years or so a minor revolution has occurred in digital imaging, one that has had a profound impact on the reproduction of photographs. Advances in technology, and in particular in computer memory, software and digital scanning, have allowed techniques that were once the domain of affluent publishing and repro houses to come into the home. Several of these advances have assisted in the preparation of this book.

Of particular relevance to this field guide has been the ability to modify and manipulate images to enhance them as aids for identification. At the simplest level, this has meant 'tidying up' images where, for example, a confusingly cluttered background detracted from a photograph's clarity. More significantly, the ability to cut out and superimpose images with relative ease has enabled far more photographs to be used per page than could have been contemplated in the past. Composite pictures have been created for certain bird groups, notably those seen most usually in flight, allowing easy species' comparisons to be made.

This brings us to the potentially thorny area of significant image manipulation. Using digital-imaging software I have altered subtly the appearance of some of the images in this book to enhance their usefulness as aids to field identification. Thus, for example, a degree of uniformity has been introduced in terms of lighting, and in a few instances key features have been enhanced where these were not clearly shown.

Computer software allowed the easy removal of debris in the water surrounding this
Wood Sandpiper.

in the original photograph. I have only undertaken this form of image manipulation on my own photographs.

None of the above imaging processes is particularly novel and many are employed, to a greater or lesser extent, in most current publishing ventures, whether or not this is spelt out. However, for this field guide, I have taken the process one stage further. Rather than make do with poor-quality images or substitute artwork for seldom photographed species, in certain instances I have actually created images from scratch using computer software.

This approach was employed only if (given the deadlines and financial constraints of this book) I was unable to take a suitable photograph of a given species or obtain it from other sources. These computer-generated images are clearly identified as such in the photo credits. They are intended to enhance the usefulness of this field guide and no attempt has been made to disguise their digital origins.

Distracting background twigs have been removed from this photograph of a Rüppell's Warbler using computer software.

If you are bothered by this approach then judge these illustrations as you would artwork rather than real photographs. They are my personal interpretation of what the bird in question looks like – I just happen to have used computer software to create them rather than paint.

Choosing the Images

Images have been selected for their relevance in the field guide context. Hence, every attempt has been made to choose photographs that portray important identification features and that capture the jizz of the species in question. Wherever possible, clearly lit images have been used, but the aesthetic qualities of the pictures were taken into consideration when designing the book.

When designing the layout of each page of images, emphasis was given to the plumage most typical of each particular bird in the context of the Mediterranean region. For most species, the dominant image is that of an adult, priority typically being given to males where plumage differences between the sexes exist. So the bird is shown in the most appropriate plumage for the region. Generally speaking, this means breeding/summer plumage, but for winter visitors non-breeding plumage is shown if this is appropriate. Wherever possible, I have tried to include as many additional plumages and poses as space will permit to increase the book's usefulness as an identification aid.

Topography

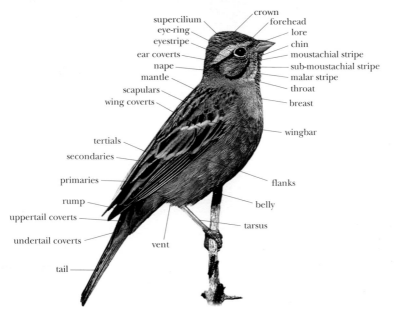

supercilium
eye-ring
eyestripe
ear coverts
nape
mantle
scapulars
wing coverts

crown
forehead
lore
chin
moustachial stripe
sub-moustachial stripe
malar stripe
throat
breast

wingbar

tertials
secondaries

primaries
rump
uppertail coverts
undertail coverts

vent

flanks
belly
tarsus

tail

Glossary

Adult A full-grown, sexually mature bird.

Axillaries Feathers that cover the 'armpit'.

Carpal (as in carpal patch) An area of feathering that typically corresponds to the underwing primary coverts.

Cere Bare skin seen in some birds at the base of the bill and surrounding the eye.

Colour phase Adults of some birds – notably certain skua species – occur in two, strikingly different colour forms, light and dark, which are often referred to as 'phases'.

Flight feathers Primary, secondary and tertial feathers on the wing that together form the bulk of the wing area.

Forewing The leading margin of the wings, often synonymous with the upperwing coverts.

Immature A young bird that has not yet reached adulthood.

Juvenile A young bird in the first year of its life.

Lore Feathered area between the base of the upper mandible and the eye.

Moult The process of shedding worn feathers and replacing them with new ones. Moulting occurs in all birds, but the pattern of moult varies from species to species.

Passage migrant A bird that passes through the region on migration both in spring (mainly April–May) and in autumn (mainly August–September).

Primaries The outermost flight feathers that form the 'hand' of the wing.

Secondaries The flight feathers on the central part of the wing.

Speculum A colourful and often iridescent patch on the secondary feathers of dabbling ducks.

Tertials The innermost flight feathers.

Underparts Feathering on the underbody and underwing.

Upperparts Feathering on the upper body (including the head), wings and tail.

Climate, Vegetation and Habitats

The term 'Mediterranean' applies not only to the eponymous sea but also to a climatic type typified by long, hot and dry summers, and mild, wet winters where frosts and snow are almost unknown. Spring and autumn are brief transition seasons of relatively unsettled weather sandwiched between the region's two dominant seasons.

The extent of the Mediterranean climate's influence corresponds roughly to the geographical range of the olive tree. It affects much of the Mediterranean coast and, depending on the topography of the land, the influence often extends inland to a distance of 100km or more. Indeed, low-lying land across the southern half of the Iberian Peninsula is essentially Mediterranean in character.

Given the size of the region, it is hardly surprising that the climate varies across the Mediterranean. In the west, the maritime influence of the Atlantic Ocean makes this the wettest area. Moving inland from the shores of the sea, the land masses of Europe to the north and Asia to the east gradually transform the weather patterns into a continental climate, where winters are much more severe. To the south, warmer, arid land significantly curtails the extent to which the Mediterranean climate is felt in North Africa and the Middle East. Needless to say, these are generalisations, and local factors – particularly the presence of mountain ranges – have a profound influence upon the climate in a given area.

Climatic factors – notably temperature and rainfall – have an enormous influence on the evolution of vegetation. Consequently, we find that Mediterranean plants are adapted to cope with the drought and heat of summer and to make the most of the mild winters. However, the appearance of the landscape has been severely modified by human activities, so much so that almost no natural climax vegetation survives. Thousands of years of human colonisation and land use, with the associated activities of tree felling, fires, grazing pressure and ploughing, have had a profound impact.

Without people, the Mediterranean would still be largely wooded. As it is, only fragments of unspoilt natural forest survive today and instead we find a mosaic of open habitats, which, as a rule, become increasingly open and barren over time. While this may sound rather depressing, human activities have opened up a wide range of niches, far more in fact than would be found if woodland alone was present. This fact is reflected in the diversity and abundance of birdlife in the region. While some of the birds covered in this book are habitat generalists, many are extremely specific. Therefore, being able to distinguish the region's different habitats aids bird identification and can also help pinpoint the best places to look for certain target species.

Off the beaten track, there are still plenty of relatively unspoilt areas of coast that provide a landfall for tired migrants in spring.

Freshwater Wetlands

Throughout the region covered by this book, the availability of freshwater bodies is always limited, both in terms of size and seasonality. Consequently, sites where fresh water occurs are of prime importance to many bird species and it follows that all forms of water body, be they large or small, are focal points for birdwatchers too.

Given the fact that water is at a premium in the Mediterranean, it is hardly surprising that most freshwater habitats have been severely affected by human activities: damming, water abstraction and watercourse diversion are all commonplace. Apart from the few locations that have protected reserve status, those wetlands that survive in the Mediterranean region today do so, for the most part, *despite* the best efforts of man and not *because* of them.

Around the Mediterranean, streams and rivers tend to be ephemeral, influenced by the seasonal precipitation that falls mainly between October and February. At times torrential during the winter months, the flow of these watercourses typically drops markedly in summer and smaller streams are generally bone dry by early July. The appearance of insect larvae and adults in early spring provides a welcome bonanza for migrant birds such as flycatchers and warblers, while the tadpoles that throng the dwindling pools and trickles of water are easy pickings for Little Bitterns and Night Herons. Later in the season, the vegetated margins of dry streams – sometimes lined with Oleander and Tamarisk – may harbour nesting birds. In the east of the region, look for breeding Lesser Grey Shrikes and Rufous Bush Robins in these locations.

Flat, coastal land often lies wet in winter. Where the ground is particularly low-lying and soil conditions permit, pools and marshes form and often retain their water into April or May. Insects with aquatic immature stages, such as dragonflies and damselflies, are often abundant and frogs and toads make use of the water for spawning. Such areas can be superb for watching migrant insect-eating birds such as pipits, wagtails, larks and pratincoles. Waterbirds such as Glossy Ibises, Squacco Herons, terns and Wood Sand-pipers take their toll of both the insects and the tadpoles. During the winter months, coastal marshes are the haunt of ducks and geese along with visiting raptors.

Even comparatively modest stretches of water will attract spring migrants such as the Marsh Sandpiper (1), Black Tern (2), Glossy Ibis (3), Ruff (4) and Squacco Heron (5).

Large permanent lakes are few and far between in the Mediterranean region but become more frequent the further north you travel away from the sea. Often fringed by extensive reedbeds and marshes, these lakes are the haunt of breeding waterbirds such as herons, egrets, cormorants and grebes, while in winter they often attract large concentrations of wildfowl. In late summer, their drying margins are alive with insects, which provide a welcome food supply for migrant waders, wagtails and pipits. Man-made reservoirs are typically more sterile than their natural counterparts. But with time, plants begin to colonise their margins and their importance to birds increases accordingly. On occasions, reservoirs are stocked with fish and, predictably, become attractive to passing Ospreys, egrets and herons.

Coastal Habitats

The warm, azure waters of the Mediterranean are the region's most important tourist attraction. It is little wonder then that most easily accessible sandy beaches – along the northern shores at least – are extremely popular with visitors and subject to widespread commercial development. Another factor limiting the Mediterranean coast's value in birdwatching terms is the sea's minimal tidal range. You do not find extensive mudflats, rocky shores or sand flats exposed at low tide as you do along the Atlantic coast of Europe. Consequently, feeding grounds for waders and wildfowl are restricted and their numbers are comparatively small. Do not despair, however, because plenty of relatively unspoilt coastal habitats still survive, with local concentrations of birds and a fair selection of specialities for good measure.

The open sea can be a frustrating place for the birdwatcher at the best of times and particularly so in the Mediterranean. The number and variety of seabirds breeding in the region is rather limited, compared to, say, the North Atlantic, and so seawatching can be a bit disappointing. However, shearwaters can be seen from rocky headlands or during ferry crossings, and careful scrutiny of Yellow-legged Gull flocks may reveal the occasional Audouin's Gull. You can expect least variety during the summer months, but watch out for migrant skuas in spring and autumn, and for seasonal visitors such as the Lesser Black-backed Gull in winter.

Flocks of White Pelicans occasionally grace large lakes in the eastern Mediterranean. Most are heading to or from the species' main nesting colonies in Romania's Danube Delta.

Because of the Mediterranean Sea's limited tidal range, sandy and pebble beaches have comparatively little available food to offer birds. On top of this, there is the problem of human disturbance. However, loafing Yellow-legged Gulls can turn up almost anywhere. Migrant waders occasionally pause on even the most popular of beaches in spring and autumn, and in winter there is always a chance of finding a Sanderling. In a turn of events, some popular beaches on the northwest coast of Majorca now boast the year-round presence of Audouin's Gulls.

Many parts of the Mediterranean coast have stunning and dramatic cliff formations. In a few, privileged areas, notably the islands of the Aegean, such cliffs are the haunt of Eleonora's Falcons. Most of these birds nest on inaccessible offshore islands, but their favoured clifftop hunting areas – for autumn migrants – are more widespread.

Where large rivers empty into the Mediterranean, silt deposition occurs, creating estuaries and mudflats. The limited tidal range means that comparatively small areas of potential feeding grounds are exposed, and hence their importance for birds is somewhat limited. However, those areas of mud that do become exposed will harbour a reasonable variety of waders. Diving terns and long-legged wading birds such as herons and egrets can exploit estuarine niches unavailable to the smaller waders. Loafing gulls are always present and on the lookout for an easy meal.

From a birdwatching perspective, the most interesting Mediterranean coastal habitats are man-made. In some areas, seawalls have been built to prevent flooding or to create pools for fish- or shrimp-rearing, and here extensive brackish channels and lagoons have formed. Even more impressive are the pools, channels and drying beds associated with the extraction of salt from seawater by evaporation. The process takes a considerable period of time and so a mosaic of pools at different stages of extraction, and hence salinity, is created. Black-winged Stilts, Avocets and Kentish Plovers are present in such areas year-round. If conditions favour them, Greater Flamingos can be locally common in winter, sometimes forming sizeable flocks. Migrant waders and terns can be abundant in spring and autumn, but again only if water depth and feeding conditions suit their needs.

The varied coastline can be good for the Mediterranean Gull (1), Grey Heron (2), Black-winged Stilt (3) and Little Egret (4). However, don't necessarily expect to find them together in the same habitats.

Woodland and Forest

Mediterranean woodland has been, and continues to be, altered radically by the region's human population. Forests, which once would have dominated the landscape, have been much reduced by several millennia of tree felling. And it is not just the extent of the tree cover that has been affected: the character and species composition has also been changed in many areas, for example by the selective planting or felling of particular tree species. Despite the often drastic impact of man, however, forested areas throughout the region retain their importance for woodland birds.

In areas where the Mediterranean climate exerts its influence, evergreen oaks typically dominate the woodland scene. Drought-resistant conifers, particularly pine species, also occur. Their presence and abundance is influenced by factors such as soil type, altitude and, inevitably, people: Stone Pines are widely planted as a source of pine nuts and other conifers are used as soil stabilisers and windbreaks. Beneath the canopy of these evergreen woodlands grows an understorey of medium-sized shrubs, which are the main component species of the scrub habitats that appear after tree clearance. Evergreen woodlands are important resting and feeding spots for migrant birds such as warblers and flycatchers. In addition, species such as the Golden Oriole, Nightingale and Hoopoe often breed in these habitats; flocks of winter-visiting finches and thrushes also find them attractive.

Where the influence of the Mediterranean climate is moderated by mountains, deciduous trees begin to dominate as altitude increases. This is particularly noticeable in the eastern half of the region and the further north you travel away from the Mediterranean proper. Breeding birdlife tends to be more varied in this habitat, and species such as the Bonelli's Warbler, European Honey Buzzard and Common Buzzard, plus various woodpeckers, are typical. During the winter months, these deciduous woodlands also host flocks of finches.

Where the mountains are high enough to have a profound influence on temperature and rainfall, deciduous trees are finally replaced by conifers; the altitude at which this transition occurs is lower the further north you travel from the Mediterranean Sea. These high-altitude forests are the haunt of specialist birds such as Crossbills; the open aspect of these woodlands makes raptor-watching a worthwhile pursuit.

Mediterranean woodlands are the haunt of residents such as the Common Buzzard (1) and Middle Spotted Woodpecker (2), as well as migrants that include the Golden Oriole (3), Common Redstart (4) and Collared Flycatcher (5).

Scrub Habitats

In the wake of woodland clearance, open scrub habitats come to dominate the Mediterranean landscape; the floral elements of these habitats have their origins as woodland understorey shrubs. Although rainfall and soil type have a huge influence on the habitat that develops, from an ecological perspective the act of tree clearance is seen to trigger an inexorable degenerative process of dwindling vegetation cover. Over time, the dense cover of large shrubs and small trees gradually gives way to low-growing plants that form an incomplete ground cover. At the two extremes of this gradation in scrub habitats are vegetation types distinct enough to merit names. Dense, shrubby plant cover is referred to as maquis, while barren, poorly vegetated habitat is called garrigue. Both support discrete and habitat-specific birds and so, for the birdwatcher, it is worthwhile learning to distinguish the two.

Maquis is recognised by the presence of evergreen shrubs that range in height between 1m and 5m. The Strawberry-tree, Tree Heath and broom relatives are dominant, and various lower-growing *Cistus* species put on colourful flowering displays in spring. Above altitudes of 600m, Kermes Oak (recognised by its tiny holly-like leaves) and Box begin to feature, and ecologists term this transition habitat as pseudomaquis. At whatever altitude they occur, maquis and pseudomaquis are favoured haunts of Woodchat Shrikes, which find the rich insect life much to their liking; European Bee-eaters often feed overhead and nest wherever exposed banks of sandy soil occur. Cirl Buntings sing their repetitive songs in spring but arguably the most characteristic songsters of the maquis (although not always the easiest to see) are the *Sylvia* warblers. Some maquis-frequenting members of this genus – Subalpine and Sardinian Warblers, for example – are widespread across the Mediterranean region, while others have extremely restricted ranges.

Maquis habitat is at its most colourful in spring, which is the time to look for Subalpine Warblers (1), Woodchat Shrikes (2) and European Bee-eaters (3).

Garrigue is characterised by the presence of dwarf shrubs, none of which exceeds a height of 1m; typically most are even shorter than this. Many of the shrubs are dense and spiny, and there is usually plenty of bare, stony ground between them. Garrigue is also spelt garigue, and in Greece this habitat goes by the name *phrygana*. Despite the paucity of vegetation, garrigue often supports a surprising wealth of seasonal insect life, notably bush-crickets. Specialised *Sylvia* warblers favour this habitat – Marmora's, Balearic and Spectacled Warblers, for example, although the first two have extremely restricted ranges. In Iberia and northwest Africa, garrigue is the favoured habitat of Thekla Larks.

Grassland, Steppe and Stony Uplands

Despite its widespread occurrence throughout the Mediterranean region, grassland is seldom the natural climax vegetation. Rather, it is usually the final stage in the degenerative process of vegetation loss that occurs after tree clearance. Severe grazing pressure and fires often contribute to the occurrence of the stony, seemingly barren grasslands so typical of many areas. Although the plant species composition may differ from true steppe habitat, the overall appearance of many Mediterranean grasslands is superficially similar to this Asiatic habitat. Created and maintained by centuries of grazing, pockets of true steppe can be found from Romania and Hungary south to Turkey and the Middle East; the habitat becomes far more widespread as you move east into Asia.

Although in biological terms Mediterranean grasslands may not seem particularly productive, they do support a surprising diversity of seasonal insects, notably bush-crickets and grasshoppers. In turn, these allow open-country birds to flourish, and larks of various species – notably Lesser Short-toed and Calandra – are locally common. In the east of the region, Isabelline Wheatears can be found where true steppe habitat occurs. Elsewhere in the Mediterranean, Northern and Black-eared Wheatears replace them, particularly on stony slopes where boulders dot the landscape.

In the west of the region, garrigue is the haunt of the Thekla Lark (1) and Spectacled Warbler (2). Across the Mediterranean, the habitat's diverse reptile fauna attracts hunting Short-toed Eagles (3).

Deserts and Semi-deserts

The arid regions of North Africa and the Middle East are challenging, both for their resident birds and for visiting birdwatchers. Searing daytime temperatures and a virtual absence of freely available water mean that only specialists can survive here. However, both sandy and stony deserts potentially harbour fascinating birds, and semi-desert (where scattered vegetation occurs) can be equally rewarding to explore.

Biological productivity in the desert is dependent upon the limited rainfall, which usually occurs during the winter months. Typically, precipitation is extremely patchy and some areas may not receive any water for several years. Consequently, desert birds are rather nomadic and their own patchy distributions usually correspond to areas where rainfall has occurred in the past few months. Find one of these spots and you should find plenty of birds. It is also worth checking any wadis (dry riverbeds) that you come across. Sufficient moisture is usually retained in the ground for communities of acacia trees to thrive. Waterholes and springs are also important focal points for desert birds.

Cream-coloured Coursers and various sandgrouse and wheatear species are among the highlights of desert birding. However, arguably the greatest diversity of arid-country species is found among the larks: Desert, Bar-tailed and Hoopoe Larks are just three of the star attractions.

Open, stony and grassy habitats support a surprising range of birds. The Lesser Short-toed Lark (1) favours open, poorly vegetated terrain, while Cretzschmar's Bunting (2) and Rock Nuthatch (3) prefer stony ground in the east of the region. The Isabelline Wheatear (4) is a bird of true steppe grassland.

Desert scenery is frequently stunning, providing a perfect backdrop for specialist birds such as the White-crowned Wheatear (1), Temminck's Lark (2), Desert Lark (3) and Blackstart (4).

Agricultural Land

In contrast to farmland in northern Europe, agricultural fields around the Mediterranean are often comparatively rich in wildlife. Although there are exceptions, of course, pesticides, herbicides and fertilisers are applied less extensively. Consequently, even fields that are currently farmed often support a profusion of wildflowers – weeds in most farmers' eyes – and insect life is varied too. Fallow and abandoned fields are generally even more productive.

A consequence of this less intensive approach to farming is that there is a relative abundance of food – notably seeds and invertebrates – for birds to eat. Resident species such as the Red-legged Partridge, Corn Bunting and Crested Lark favour open, arable fields and are joined by summer visitors such as Black-headed Buntings in the east of the region. Outside the breeding season, flocks of larks, buntings and finches favour this habitat. Olive groves and orchards of fruit trees contribute to the diversity of this habitat and their presence encourages breeding species such as the Little Owl, Hoopoe, Woodchat Shrike and Orphean Warbler.

Villages and Gardens

Most Mediterranean villages still retain a rural outlook on life, and as a result the inhabitants make good use of the land surrounding their houses to grow crops. Even the most humble of dwellings is likely to boast a few grapevines, fig and citrus trees, and rows of potatoes and other vegetables, all nurtured and watered by their diligent owners.

For the most part, the owners of these smallholdings are indifferent to the presence of birds in their gardens. Depending on which part of the Mediterranean you visit, you should have no trouble finding European Serins, European Goldfinches and sparrows (either House or European Tree) throughout the year; swallows, House Martins and swifts are summer visitors to the region and use roofs and outbuildings for nesting. Many villages also boast a pair or two of Scops Owls. The cryptic plumage of this summer visitor makes it difficult to find but its presence is comparatively easy to detect after dark, when the bird utters its peculiar sonar-blip call.

Depending on when and where you visit, colourful arable fields may harbour Rollers (1), Black-headed Buntings (2), Lesser Kestrels (3) and Stone-curlews (4).

Olive groves and gardens harbour European Tree Sparrows (1), Scops Owls (2), Crested Larks (3) and Rose-coloured Starlings (4), although not all occur in the same area or in the same season.

Migration

Although the Mediterranean boasts a wealth of bird species that are resident or that make comparatively short seasonal or altitudinal movements within the region, a large proportion of the breeding species are seasonal migrant visitors. Furthermore, the Mediterranean is on the flyway for passage migrants heading to and from more northerly latitudes in Europe, and their appearance enlivens the birdwatching scene tremendously. You never know what might turn up, and you may even be treated to the spectacle of numbers, with bushes full of warblers, or flocks of storks and raptors floating overhead. For most birdwatchers, migration is what makes the Mediterranean one of the most exciting destinations in the world.

The significance of the Mediterranean to migratory birds is rooted in the geography of the region and in the birds' patterns of migration. Although a few summer migrant visitors winter in Southeast Asia, most have wintering grounds in Africa. Consequently, for birds heading north in spring or south in autumn, the Mediterranean Sea is a significant obstacle in their path, particularly for those species with African wintering quarters.

Large, diurnal migrants, such as raptors and storks, rely on land-generated thermals to assist their passage and consequently are reluctant to migrate over the open sea. As a result, huge concentrations of these species funnel through bottlenecks where the sea crossing is reduced to a minimum; the Strait of Gibraltar and the Bosporus at Istanbul are well known for this phenomenon.

Wherever possible, smaller migrant species, many of which fly at night, also try to avoid long sea crossings. They tend to hug the coastline of the Mediterranean as much as possible, but at some point most have little option but to head out to sea and hope for the best. Spring and autumn migrations coincide with periods of unpredictable and unsettled weather. Consequently, many migrants – and especially nocturnal ones – become grounded by adverse winds or heavy rain, either at their point of departure or at the first landfall they make after crossing open water. However, what is bad news for the birds is good news for the birdwatcher: on such occasions, Red-backed Shrikes, Red-throated Pipits, Whinchats and various flycatcher and warbler species can be locally abundant. The spectacle is usually short-lived, however, because once the weather clears up the birds begin to disperse, resuming their migration after dark.

White Pelicans, and certain other large daytime migrants, typically fly in V-formation and make a stunning sight.

19

Spring migration in the Mediterranean region often starts as early as the beginning of March with the appearance of swallows, House and Sand Martins, and Yellow and White Wagtails. Good numbers of White Storks are also on the move by the middle of the month, but it is not until April that the numbers and variety of passerines and waders really build up. Migration across the western half of the Mediterranean tends to be a week or two more advanced than in the eastern half. So, on Majorca, for example, you might expect the maximum number and variety of migrants to occur in the middle two weeks of April. In Greece, late April and early May are generally the peak times. Although spring migration as a whole spans a period of three months or so, the peak period for any given species is usually comparatively brief. Sometimes this can be a matter of days, but it is seldom more than a week.

Compared to spring migration, that which occurs in autumn is often more difficult to discern on the ground. The period during which any given species migrates tends to be less concentrated and individuals often tend to linger around the coast longer than they would do in spring. As a general rule, adult birds start their southerly migration before juveniles of the species, appearing from mid-August onwards. The migration of storks and large raptors extends from late August throughout September, and the passage of immature passerines continues well into October.

It is important to remember that the arrival in spring, and departure in autumn, of breeding species is not the only form of migration that occurs in the Mediterranean. The region also hosts large numbers of wintering birds that nest further north in Europe. So, autumn sees an influx of thrushes, finches and buntings, which depart again in early spring.

Seasoned visitors to the Mediterranean will tell you that no two years are the same when it comes to bird migration. The peak appearance of any given species can vary by as much as two weeks from year to year. And sometimes certain species fail to turn up at migration hotspots in any significant numbers at all when in past years they may have been abundant. Any number of reasons could be behind this variability. For example, prevailing weather conditions may have dictated alternative migration routes, or perhaps there was no adverse weather to ground the migrating birds. You never know for sure what is going to happen during migration and this is half the fun for most birdwatchers.

Numbers of Rose-coloured Starlings reaching the eastern Mediterranean varies from year to year. However, in 2002 the influx was huge, involving tens, if not hundreds, of thousands of birds.

Visiting the Region

The Mediterranean is Europe's premier holiday destination. The northern coastline in particular is littered with resorts and visiting the region has never been easier. Although visitor numbers and development have eroded the wildlife value of many popular spots, most holidaymakers seldom stray far from the nearest safe beach. Consequently, prime birdwatching locations can usually be explored in solitude.

Nowadays it is easy to arrange your own holiday accommodation and car hire. Furthermore, birding guides exist for almost every Mediterranean destination, providing details of where to go and what to look for. Consequently, most birders to the region arrange their own self-guided tours. Almost any Mediterranean destination is likely to repay a visit, but I would urge you *not* to visit those areas – notably Malta and Cyprus – where migrant birds are massacred in their millions each year. Until conservation laws are enforced here, vote with your wallet by going elsewhere and protest loudly against the slaughter whenever possible. On a more positive, and personal, note there are four locations – two in the Mediterranean proper and two on the fringes – that I feel merit being singled out as places you *must* visit.

Heading the list is the Greek island of Lésvos. The island boasts typical eastern Mediterranean species plus a selection of birds more characteristic of neighbouring Turkey. Visible migration can be spectacular.

Between southern Peloponnese and Crete lies the Greek island of Antikythera, an important staging post for migrant birds. It also boasts a significant percentage of the world's breeding population of Eleonora's Falcons. The Hellenic Ornithological Society, with help from BirdLife International, plans to open a bird observatory on the island, so monitor the press and internet for details.

Although not strictly speaking part of the Mediterranean proper, southern Israel is a superb spot for observing migrant birds heading to and from this region. Bird of prey and stork passage can be spectacular, and any shade or water acts as a magnet for migrant passerines. Eilat, the traditional destination of choice, is blighted by modern development, so consider a kibbutz-based visit as an alternative.

This book's distribution maps extend north to the Black Sea and include Romania. The climate here may be more continental than Mediterranean, but the birdlife in the east of the country has much in common with areas further south. The Danube Delta remains the most superb and extensive wetland in Europe, while the neighbouring Dobrogea region boasts European Bee-eaters, Rollers and Red-footed Falcons in abundance, along with Asiatic species such as the Pied Wheatear and Rose-coloured Starling.

For more information about visiting these regions, *see* page 187.

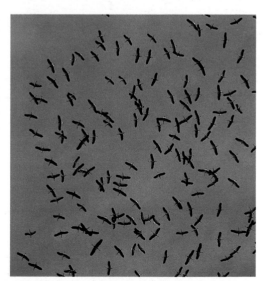

Large, diurnal migrants such as these White Storks typically use thermals to give them 'lift' and assist their passage. This flock was photographed near Israel's Mediterranean coast.

21

1. GREAT CRESTED GREBE *Podiceps cristatus* 46–51cm

Graceful, slender-necked waterbird with grey-brown upperparts and pale underparts. In breeding plumage, has orange-buff ruff, white cheeks and a black cap and crest; plumage less ornamental at other times. Bill is dagger-like and pink. Dives in search of fish. Nests on floating mound of vegetation. Found on lakes in breeding season; in winter also on coasts. Local in summer but widespread in winter.

2. RED-NECKED GREBE *Podiceps grisegena* 40–45cm

Similar to Great Crested but smaller and more compact. In breeding season has red neck and greyish-white cheek contrasting with black crown; plumage otherwise reddish brown. In winter, neck looks grubby reddish grey; contrast between dark crown and greyish cheeks is less marked. At all times bill is yellow at base and dark-tipped. Breeds very locally in E of region; more widespread and typically coastal in winter.

3. BLACK-NECKED GREBE *Podiceps nigricollis* 28–34cm

Waterbird with diagnostic head profile comprising uptilted black bill and steep forehead. In breeding plumage, shows black head, neck and back with yellow ear tufts. In winter, looks black and white. Red eye seen at all times. In flight, all birds show white on trailing edge of inner wing. Nests on shallow lakes. In winter, also seen on coasts. Local in summer; widespread in winter.

4. SLAVONIAN GREBE *Podiceps auritus* 31–37cm

Also called Horned Grebe. Similar to Black-necked but has flat-topped head and pale-tipped bill that is symmetrical in profile. In breeding plumage (seldom seen in region), has reddish neck and flanks with black cheeks and yellow ear tufts. In winter, looks black and white with dark patch on otherwise white cheeks. Scarce on sheltered coasts (especially Adriatic), Sept–Mar.

5. LITTLE GREBE *Tachybaptus ruficollis* 25–29cm

Tiny waterbird. In breeding plumage, shows chestnut on cheeks and neck and green spot at base of bill; otherwise dark brown except for pale powderpuff of feathers at rear end. In winter, has brown upperparts and buff underparts. Builds floating nest of plants. Swims buoyantly and dives well. Found on lakes and slow-flowing rivers. In winter, also occurs on sheltered coasts. Locally common across region.

6. RED-THROATED DIVER *Gavia stellata* 55–65cm

Elegant waterbird. Swims low in the water and dives frequently. Neck slender; head and narrow, dagger-like bill held characteristically tilted up at an angle. In non-breeding plumage (most often seen in region), has grey upperparts speckled with white dots, and white underparts. In breeding plumage (sometimes seen in late winter) acquires red throat and grey neck. Winter visitor (Oct–Mar) to coastal waters.

7. BLACK-THROATED DIVER *Gavia arctica* 65–75cm

Robust waterbird that dives well and frequently. Head and neck thickset. In winter plumage (most often seen in region) has grey-brown upperparts and whitish underparts. Body looks mainly dark in water but sometimes a white flank patch is visible. In breeding plumage (sometimes seen in late winter in region), acquires unmistakable black, white and grey markings. Winter visitor (Oct–Mar) to coastal waters, mainly Adriatic to Aegean; also found in Black Sea.

Birds of the
Mediterranean

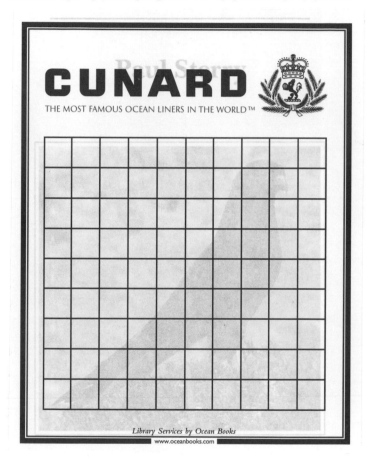

CHRISTOPHER HELM
LONDON

Published 2004 by Christopher Helm,
an imprint of A & C Black Publishers Ltd
37 Soho Square
London W1D 3QZ

ISBN 0-7136-6349-9

A CIP catalogue record for this book is available from the British Library

A & C Black uses paper produced with elemental chlorine-free pulp,
harvested from managed sustainable forests.

www.acblack.com

Most of the maps were originally researched and drawn by Mark
Beaman, and appeared previously in *The Handbook of Bird Identification* by
Mark Beaman and Steve Madge (Christopher Helm, 1998). New maps
have been drawn, or where necessary existing maps have been updated,
by Carte Blanche, Paignton, Devon.

Edited and designed by D & N Publishing
Lowesden Business Park, Hungerford, Berkshire.

Colour origination by Nature Photographers Ltd.

Printed and bound by Times Offset (M) Sdn. Bhd.

10 9 8 7 6 5 4 3 2 1

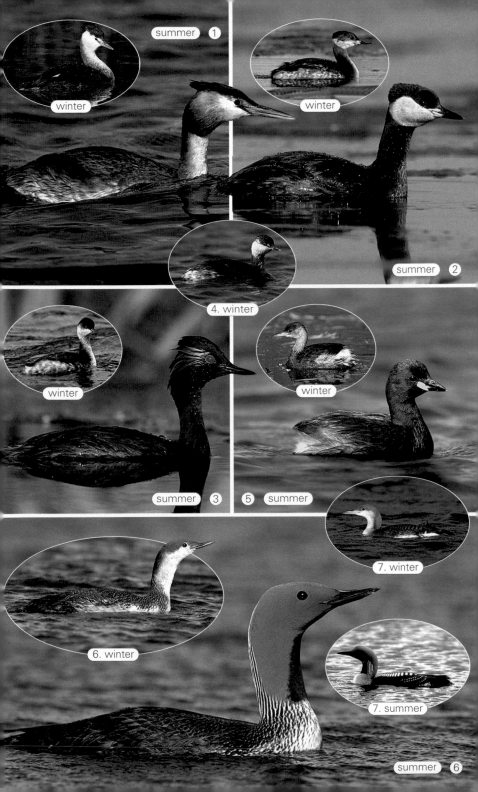

summer 1

winter

winter

summer 2

4. winter

winter

summer 3

winter

5 summer

7. winter

6. winter

7. summer

summer 6

1. CORY'S SHEARWATER *Calonectris diomedea* WS 100–125cm

Long-winged and powerfully built seabird. Upperparts, including head, are brown, with darker wingtips; underparts (except for flight feathers) whitish. Black-tipped yellow bill markings are visible only at close range. Typically glides on wings that are stiffly held and slightly bowed. In windy conditions, often rises high between lengthy glides. Often seen in sizeable groups at sea; observed from coastal headlands and ferries. Widespread summer visitor.

2. MEDITERRANEAN (YELKOUAN) SHEARWATER
Puffinus yelkouan WS 80–89cm

Smaller than Cory's Shearwater and with proportionately shorter wings. Body is cigar-shaped; at close range feet are seen to project beyond tail. Upperparts sooty-brown. Undertail coverts dark; otherwise underparts are whitish in subspecies *P. y. yelkouan* from E Mediterranean but dusky in subspecies *P. y. mauretanicus* from W; some authorities consider these races to be separate species. Banks and glides on stiffly held wings in long lines low over the sea. Sometimes feeds in groups surprisingly close inshore. Widespread and common.

3. EUROPEAN STORM-PETREL *Hydrobates pelagicus*
WS 36–39cm

Tiny seabird that resembles a House Martin from a distance, appearing mainly dark but with diagnostic white rump. Flight is powerful for a bird of this size, typically comprising direct flight interspersed with glides and periods when feet are pattered on surface of sea. Breeds on isolated islands in rock crevices and burrows; only visits colonies after dark and so breeding status difficult to assess. Seldom seen close to land and difficult to observe. Widespread summer visitor.

4. NORTHERN GANNET *Morus bassanus* WS 165–180cm

Distinctive seabird, unmistakable when seen well. Has cigar-shaped body and dagger-like bill. Adult has white plumage with black wingtips. Juvenile is dark brown, speckled with white; acquires adult plumage over several years. Glides well, but in direct flight employs deep wingbeats. Plunge-dives after fish. Winter visitor, most widespreaed in W Mediterranean.

5. DALMATIAN PELICAN *Pelecanus crispus* WS 280–325cm

Massive, whitish waterbird. In good light, looks pale blue-grey. Head has shaggy mane of back-curled feathers. Bill long, with large throat sac; orange-yellow briefly in breeding season but pinkish at other times. Seen from below in flight, adult has uniformly pale underwings; seen from above, plumage is white except for black primary feathers. Juvenile is uniformly grey-brown. Swims well. A superb flier, soars effortlessly. Rare and local at breeding lakes (N Greece and Turkey, northwards to Bulgaria and Romania's Danube Delta) May–Aug. Often coastal at other times; some linger throughout winter months.

6. GREAT WHITE PELICAN *Pelecanus onocrotalus*
WS 275–300cm

Massive, mainly white waterbird; head and neck flushed pinkish in breeding season. Bill long, with orange-yellow throat sac. Legs pinkish orange. In flight and seen from below, black flight feathers contrast with otherwise gleaming white plumage. Juvenile has brownish plumage and a yellow throat sac. Soars and glides effortlessly; often seen in large flocks on migration. Locally common at breeding lakes (N Greece and Turkey, northwards to Romania's Danube Delta) May–Aug; some linger into autumn, but winters in Africa. Seen on migration in E of region.

1. PYGMY CORMORANT *Phalacrocorax pygmeus* 45–55cm

Relatively small cormorant with proportionately long tail. Appears all dark at a distance. At close range, breeding adult has brownish head with white flecks on face and neck. Non-breeding adult loses white flecks and acquires pale throat. Swims low in water and can resemble a grebe when neck is stretched. In flight, can resemble a Coot, the long tail recalling that species' trailing legs. Often perches on waterside branches with wings outstretched and usually encountered in small groups. Favours mainly freshwater and breeds colonially, often nesting in trees. Main range is from N Greece and Turkey to Romania's Danube Delta; sometimes wanders in winter.

2. EUROPEAN SHAG *Phalacrocorax aristotelis* 65–80cm

Similar to cormorant in appearance and habits but smaller and with more delicate bill. Plumage shows green, oily sheen, most obvious during summer months. Breeding adult has crest and yellow base to bill; both features less obvious in winter. Juvenile is brown above and paler below. Swims low in water and dives well in search of fish. Often stands on rocks with wings outstretched. Flies with outstretched neck. Favours rocky coasts and almost exclusively marine in its habitat preference. Nests locally throughout most of the region; more widespread in coastal waters in winter.

3. GREAT CORMORANT *Phalacrocorax carbo* 80–100cm

Large and bulky waterbird with long, hook-tipped bill. Often appears all dark but, at close range, back looks scaly owing to pale feather margins and plumage has oily sheen. In summer, adults across most of Europe have white on face and thighs; summer adults from NW Africa have white feathering extending down neck to upper breast. In winter, most adult birds look rather uniformly dark: white thigh patch is lost and face appears grubby. Juvenile has grey-brown upperparts, streaked pale throat and neck, and whitish underparts. Swims low in water and dives well in search of fish. Often stands on posts with wings outstretched. Flies with outstretched, slightly kinked neck. Breeds locally, typically colonially in trees and usually beside freshwater. Widespread in winter, when it occasionally favours coastal seas as well as freshwater.

4. GREATER FLAMINGO *Phoenicopterus ruber* 125–145cm

Unmistakable and elegant long-legged and long-necked waterbird. Adult's pale pink plumage can look almost white in poor light. Pink, banana-shaped bill has black tip. Legs long and pinkish-red. In flight, wings show black flight feathers and reddish-pink coverts; neck held outstretched with legs trailing. Juvenile is grey-brown with dull legs and bill. Long neck is typically held in 'S' shape. However, with resting birds, head and neck are often pressed close to body; feeding birds extend neck so that upturned bill can be used to filter food from water. Seldom seen alone and often encountered in large flocks. Favours brackish lagoons and saltpans, but only where food items, such as brine shrimps, are abundant. Regular breeding locations include Camargue, S Iberia, N Africa and Turkey, but even here nesting is still somewhat unpredictable. Only attempts to nest if water conditions suit both feeding and nest-building and will abandon previously favoured locations if conditions are not suitable. More widespread in winter than in summer, but extreme habitat specificity means that flocks are still always local in occurrence.

1

2

summer

imm

juv

summer

4

4

3

1. WHITE STORK *Ciconia ciconia* 100–115cm

Large and unmistakable bird. Adult plumage is grubby white except for black flight feathers. Bill long, dagger-like and red; long legs are pinkish red. Juvenile similar to adult but colours of bill and legs are duller. Walks in slow and deliberate manner, using bill to stab at fish, amphibians and large insects. Occasionally scavenges at rubbish dumps. In flight, all birds look white with contrasting black flight feathers; legs and neck are held outstretched. Capable of sustained soaring and gliding. Migrating flocks rise on thermals to great heights; migration into and out of Europe is mainly via Gibraltar and Bosporus. Nests on roofs of houses, churches, telegraph poles etc, and feeds on nearby marshes and fields. Usually tolerant of human presence. Breeds mainly Iberia and NW Africa in the W and Greece and Turkey northwards in the E. Present mainly Apr–Sept.

2. BLACK STORK *Ciconia nigra* 95–100cm

Slightly smaller than White Stork. Adult looks mostly black with contrasting white underparts; at close range, upperparts show a metallic sheen. Bill long, red and dagger-like; legs long and red. Red skin around eye visible at close range. Juvenile similar to adult but black elements of plumage have brown tinge and legs and bill are dull greenish pink. In flight, all birds look dark except for contrasting white belly and chest. Scarce passage migrant, easiest to observe on migration in E Mediterranean. Breeds mainly in NE of region (present mainly May–Aug) and very locally in S Iberia (where some birds remain year round). Nests in remote wooded valleys and hunts amphibians and insects in wetlands and fields. Usually wary and unapproachable.

3. EURASIAN SPOONBILL *Platalea leucorodia* 80–90cm

Unmistakable when seen well. Adult has white plumage and unique bill that is black and flattened with a spoon-shaped tip; bill often concealed in roosting birds. Breeding birds show yellow flush to breast and at base of bill. Legs long and black. Juvenile similar to adult but bill and legs are dull pink. In flight, all birds have long, bowed wings with head and neck held extended and trailing legs. Wings of adult pure white but those of juvenile are dark-tipped. Favours shallow, coastal lagoons and feeds by sweeping bill from side to side. Extremely local breeder, but more widespread in winter.

4. NORTHERN BALD IBIS *Geronticus eremita* 70–75cm

Distinctive bird, adult having bald red face and shaggy nape feathers. Plumage blackish; metallic sheen seen in good light. Bill long, reddish and downcurved; legs red. Juvenile lacks adult's bare-part colours and nape feathers. In flight, all birds show broad wings; legs do not project beyond tail. Nests on cliffs. Feeds in fields and semi-deserts. Rare and close to extinction. Now only easily observed in Morocco.

5. GLOSSY IBIS *Plegadis falcinellus* 55–65cm

Distinctive waterbird with heron-like proportions but Eurasian Curlew-like bill. Adult can appear black but good light reveals plumage to be deep maroon with metallic sheen to feathers of wings and back. Breeding birds have white lines from base of bill to eye. In flight, head and neck are held outstretched with legs trailing beyond tail. Favours wetlands, and present in region mainly Apr–Sept. Widespread passage migrant, especially in E Mediterranean but also breeds locally across region. Nests colonially and usually feeds in small groups. Overwinters in small numbers.

1. SQUACCO HERON *Ardeola ralloides* 45–47cm

Compact and stocky heron. In breeding plumage, looks buffish-brown with streaking on crown and trailing plumes on nape; underparts white. In flight, transformed by pure white wings. Non-breeding adult and juvenile have streaked, dull brown plumage but similar wing colour to summer adult. Bill yellowish green (bluish briefly at height of breeding season) with black tip. Legs reddish in summer adult but otherwise dull yellowish-orange. Favours well-vegetated wetlands for nesting but often found on surprisingly small watercourses and in open settings on migration. Feeds mainly on amphibians and small fish. Widespread passage migrant and locally common breeding species; present mainly Apr–Sept but some individuals linger into late autumn.

2. LITTLE BITTERN *Ixobrychus minutus* 35–38cm

Smallest heron in Europe. Adult male has greyish face, black cap, back and flight feathers, and orange-buff underparts, flushed with lilac on the face. Wings show a pale panel that grades from orange-buff to greyish white. Female has similar markings to male but plumage colours are subdued and less contrasting. Juvenile has streaked brown plumage. Breeds in extensive reedbeds and seen mainly in jerky, low-level flight. Clambers with ease through dense waterside vegetation. Adopts an upstretched posture when alarmed. Feeds mainly on amphibians, including tadpoles, and small fish. Often relatively indifferent to people when on migration but otherwise difficult to observe. Migrant visitor, which occurs locally throughout the region wherever suitable, undisturbed wetland habitats occur. Often found on surprisingly small streams and ditches on migration. Present in region mainly Apr–Sept.

3. BLACK-CROWNED NIGHT HERON *Nycticorax nycticorax* 60–65cm

Distinctive and compact heron. Has hunchbacked appearance at rest and proportionately large head. Adult has black bill, crown and back, grey wings, and a pale face and underparts; eyes are large and red. Trailing head plumes acquired in breeding season. Legs yellowish. Juvenile has essentially brown plumage, heavily marked with white spots on upperparts; bare-part colours are duller than in adult. Often difficult to observe since mainly nocturnal and roosts in dense cover. Typically leaves roost site at dusk to feed on fish and amphibians. Favours extensive wetlands when nesting but often found alongside comparatively small watercourses on migration. Widespread migrant visitor to the region and locally common, mainly Apr–Sept.

4. GREAT BITTERN *Botaurus stellaris* 70–81cm

Bulky and robust heron-like bird. Streaked brown plumage affords it excellent camouflage in wetland habitats. Back, wings and underparts mainly buffish with dark brown streaks. Neck is orange-buff and streaked, and head shows dark crown and dark malar stripe. Bill is dagger-like and pinkish. In flight, broad wings and deep wingbeats recall an owl. Adopts an upstretched posture when alarmed. Territorial male utters deep booming display call in breeding season. Favours extensive reedbeds for nesting. Outside breeding season, sometimes found in surprisingly confined wetland habitats. Despite size, always rather difficult to observe given its predilection for dense wetland vegetation. Feeds mainly on fish and amphibians. Local resident of large, undisturbed wetlands along European shores of Mediterranean. More widespread during winter months thanks to influx of birds from further N in Europe.

1. GREY HERON *Ardea cinerea* 90–98cm

Large and elegant waterbird with long legs, a long neck and a dagger-like bill. Back and wings are blue-grey. Adult has black crest feathers, but head, neck and underparts are otherwise whitish except for black streaks on front of neck and breast. Juvenile is similar to adult but markings are less distinct and plumage is generally grubby. Bill and legs are yellowish, the colour brightest in adult birds. In flight, wings are broad and rounded with black flight feathers; flight pattern comprises slow, flapping wingbeats. Utters loud *frank* call. Widespread and locally common resident of wetlands; also on sheltered coasts, mainly as a non-breeding visitor, Sept–Mar.

2. PURPLE HERON *Ardea purpurea* 78–90cm

Smaller than Grey Heron with more slender head and neck. Adult looks mainly greyish-purple. Head and neck are orange-buff with black stripe running length of neck on both sides. Shows head plumes and long, streaked breast feathers. Juvenile is similar to adult but plumage is more uniformly brown. In flight, wings are broad and rounded; upperwing is purplish brown with dark flight feathers, underwing is grey with maroon leading edge. Neck is held in snake-like curve in flight and hind toe is cocked upwards. Nests in extensive and undisturbed reedbeds. Rather secretive and difficult to observe except in flight; less so on migration. Widespread summer visitor, present mainly Apr–Aug.

3. GREAT EGRET *Egretta alba* 85–100cm

The largest white heron-like bird in the region. Distinguished from Little Egret by larger size and proportionately bigger bill, which is black in breeding season but otherwise yellow. Legs are yellow in breeding season but dark at other times. Very local during summer months; colonial nesting locations include the Camargue, N Italy, N Greece and Danube Delta (breeding range may be expanding). Present at breeding colonies mainly Mar–Sept. Disperses outside this period and autumn numbers boosted by arrival of visitors from further N and E; generally widespread in E Mediterranean in winter.

4. CATTLE EGRET *Bubulcus ibis* 48–52cm

Stocky heron-like bird with essentially white plumage and diagnostic bulging-throat appearance. During breeding season, crown and back acquire an orange-buff tinge. Legs pinkish in breeding season but otherwise dull green. Dagger-like bill is orange-yellow but becomes red and pale-tipped briefly at start of breeding season. Favours dry habitats and often associated with grazing animals, which it follows to catch the insects they disturb. In flight, wings are broad and round, and the neck is held hunched up. Locally common in Iberia, the Camargue, NW Africa and Middle East; occasional elsewhere but range may be expanding.

5. LITTLE EGRET *Egretta garzetta* 55–65cm

The commonest and most widespread white heron-like bird in the region. Bill is long and dark and legs are black but with diagnostic yellow feet. Adult has trailing head plumes in breeding season. In flight, wings are broad and round. Breeds locally in colonies on extensive wetlands; much more widespread during migration times and in winter. A partial migrant, most numerous Mar–Sept, but can be observed in the region throughout the year.

juv

1

2

juv

3

summer

winter 4

4. summer

summer 5

1. WHOOPER SWAN *Cygnus cygnus* 145–160cm

Similar to Mute Swan. Best distinguishing feature is the bill, which is triangular in profile, following the line of the head; in adult, it is black-tipped and bright yellow. Juvenile has grubby pinkish-buff plumage and a pale pink bill; bill shape aids distinction from juvenile Mute Swan. All birds typically hold neck straight when swimming. Flies on broad, powerful wings and utters loud honking calls. Regular but scarce winter visitor (Oct–Mar) to wetlands of N Greece, NW Turkey and Black Sea coasts. Occurs elsewhere in harsh winters.

2. MUTE SWAN *Cygnus olor* 130–150cm

Large waterbird with long neck, held upright or in graceful S-shaped curve. Adult is mainly pure white although some acquire a faint buff-ish tinge to crown and neck. Legs black and bill reddish orange, black at the base and with a black knob. Juvenile is grubby pinkish buff with a dull pink bill. Flies on broad, powerful wings with neck outstretched. Sociable, and favours large wetlands; sometimes coastal in winter. Grazes vegetation and may upend to feed. Scarce local resident in S France, N Greece and Turkey; more widespread in winter.

3. GREYLAG GOOSE *Anser anser* 75–90cm

Robust waterbird. The region's most widespread grey goose. Adult is grey-brown with barring on back and belly produced by pale feather margins. Stern is white and neck is marked with dark, wavy feather edges. Most adults seen in the region have pinkish-orange bills; those from E are pink. Legs are pink in all birds. Juvenile similar to adult but duller and more barred. In flight, all birds show striking pale blue-grey leading edge to inner wing, on both upper and lower surfaces. Honking calls uttered by flocks. Favours extensive and undisturbed wetlands. Local resident in N Greece, NW Turkey and Black Sea coasts. Also a widespread and locally common winter visitor, present Nov–Mar.

4. BEAN GOOSE *Anser fabalis* 70–85cm

Resembles Greylag but adult is browner and head and neck look particularly dark. Bill is dark with orange patch near tip (extent of colour varies according to race). Legs are orange (pink in Greylag). Juvenile similar to adult but colours on bill and legs are duller. In flight, all birds look uniformly dark-winged. Favours arable fields. Regular but local winter visitor (Oct–Mar); sites include Camargue, N Italy and Danube Delta. Occasional elsewhere in harsh winters.

5. GREATER WHITE-FRONTED GOOSE *Anser albifrons* 65–75cm

Distinctive grey goose with brown, barred plumage, darkest on head, neck and back. Stern is white and belly has variable, thick black bars. Legs are dull orange. Bill is pink or orange (depending on race). Adults show striking white blaze on forehead (absent in juveniles). In flight, looks dark-winged and utters loud honks. Favours arable fields. Local winter visitor (Oct–Mar); regular sites include Camargue, N Greece, W Turkey and Black Sea coasts.

6. LESSER WHITE-FRONTED GOOSE *Anser erythropus* 55–65cm

Similar to Greater White-front but smaller and more compact. Adult has proportionately smaller bill, white blaze extending onto crown above eye, and yellow orbital ring; juvenile lacks white blaze but has dull yellow orbital ring. Rare winter visitor (Oct–Mar) to E of region, particularly to N Greece and Danube Delta. Found among flocks of Greater White-fronts by careful searching.

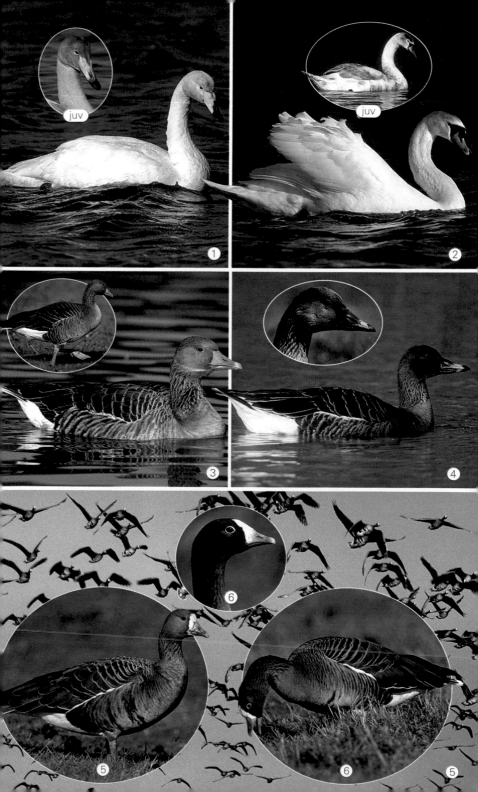

juv

juv

1

2

3

4

5

6

6

5

1. RED-BREASTED GOOSE *Branta ruficollis* 55–60cm

Small but unmistakable goose. Adult has striking black and white pattern on body and bold red markings on face and neck, these defined and separated from black markings by white lines. Has small round white patch at base of proportionately small black bill. Legs black and rather short. Juvenile similar to adult but has more barring on back and duller red colour. High-pitched calls are usually drowned out by those of grey geese with which it invariably associates in the region. Favours arable fields. Endangered. Most of the world population winters from Romania's Danube Delta south along Black Sea coast; small numbers reach N Greece, especially in harsh winters. Lone birds, as well as flocks, invariably associate with flocks of other geese, particularly Greater White-fronts.

2. COMMON SHELDUCK *Tadorna tadorna* 58–65cm

Strikingly marked goose-sized duck. Adult has dark green head and upper neck (colour can look black in some lights). Plumage is otherwise mostly white except for orange-chestnut chest band and black flight feathers on wings. Legs are pinkish red and bill is bright red; male has knob at base of bill. In flight, looks very black and white. Juvenile has white and brown plumage with pattern corresponding to that of adult. Male utters high-pitched whistling display call in spring; female utters a nasal *gagagaga*. Favours coastal habitats including estuaries and mudflats, but also occurs beside large freshwater lakes. Scarce resident breeder, mainly coastal Iberia, S France and N Greece; typically nests in burrows. More widespread across the region in winter.

3. RUDDY SHELDUCK *Tadorna ferruginea* 61–67cm

Robust goose-sized duck. Adult has orange-brown body with clear demarcation from paler buff head and upper neck. In breeding season, male has narrow black collar separating these two plumage colours. Bill and eye of both sexes is dark. Standing birds show black wingtips. In flight, wings look strikingly black and white. Juvenile is similar to adult but colours are duller. Utters nasal honking call. Favours coastal wetlands and river deltas mainly in E Mediterranean; breeds very locally (and partly resident) in N Greece, Turkey but also NW Africa, using burrows and tree holes. Dispersive and more widespread in winter.

4. MALLARD *Anas platyrhynchos* 50–65cm

Widespread and familiar duck. Male has yellow bill and green shiny head and neck separated from chestnut breast by white collar. Plumage is otherwise grey-brown except for black stern and white tail. Female has orange bill and mottled brown plumage. In flight, both sexes have blue and white speculum. Female utters familiar quacking call. Found on almost any unpolluted freshwater habitat and also on coasts in winter. Widespread year-round and commonest in winter owing to the influx of birds from N Europe.

5. EURASIAN WIGEON *Anas penelope* 45–51cm

Distinctive and attractive duck. Male has orange-red head with yellow forehead, pinkish breast and otherwise finely marked, grey plumage. Shows characteristic black and white stern and white wing patch in flight. Female has mottled reddish-brown plumage with darker feathering around the eye; best identified by association with males. Widespread winter visitor, present Oct–Mar. Favours saltmarshes and coastal grassland and invariably found in flocks. Presence often first detected by male's distinctive *wheeoo* call.

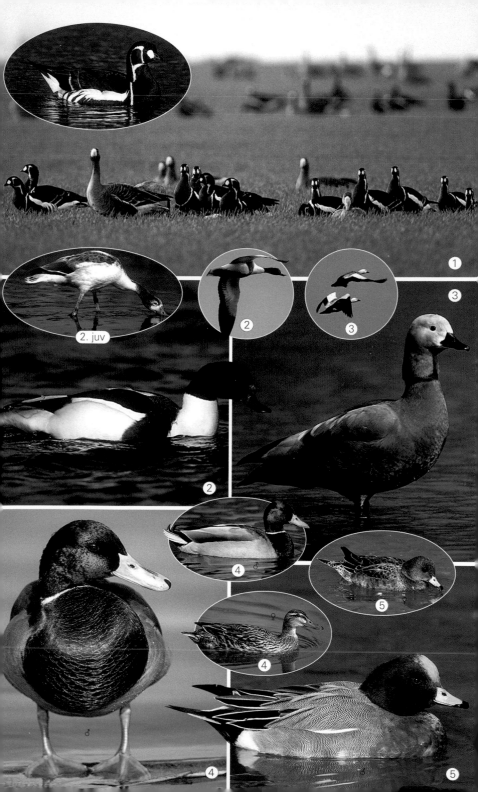

2. juv

2

3

1

3

2

3

4 ♂

5

4 ♂

4 ♀

5

♂

5

1. NORTHERN SHOVELER *Anas clypeata* 44–52cm

Distinctive duck, both sexes of which can be recognised by the long, flattened bill. Male is boldly marked with a green head, black and white patterns on the body, and reddish-chestnut flanks. Female is mottled brown and seldom seen outside the company of a male. Both sexes have a green speculum and a pale blue panel on the forewing. Favours well-vegetated wetlands. Local resident but best known as a non-breeding visitor to the region, present Oct–Mar.

2. GADWALL *Anas strepera* 45–56cm

A rather nondescript duck, but close views of male in particular reveal subtle and delicate markings. Male has grey-brown plumage with intricate dark markings and a diagnostic black stern; bill is black. Female is similar to female Mallard, with mottled brown plumage and yellow bill; best identified by association with males. In flight, both sexes show white on speculum. Male has a croaking call and female utters a mallard-like *quack*. Favours wetlands with open water. Local resident but widespread non-breeding visitor, present Oct–Mar.

3. GARGANEY *Anas querquedula* 37–41cm

Distinctive male has a reddish-brown head and a broad white stripe above and behind eye; neck and breast are brown, stern is mottled brown, but plumage is otherwise greyish. Mottled brown female is similar to female Teal; best identified by proportionately larger, all-grey bill or by association with male. Both sexes have blue forewing panel and greenish speculum but colours most striking in male. Favours wetlands and coastal marshes. Common passage migrant throughout (mainly Mar–May and Aug–Oct); small numbers breed.

4. COMMON TEAL *Anas crecca* 34–38cm

The smallest duck in the region. Male has a chestnut-orange head with a yellow-bordered green patch through the eye; plumage otherwise finely marked grey except for the black-bordered yellow stern and a striking white stripe along the upper flanks. Female is mottled and streaked brown; bill is often flushed orange-yellow at base. Both sexes show a green and black speculum, defined by a white wingbar and white trailing edge to wing. Male has a whistling call and female utters a soft *quack*. Favours coastal wetlands and saltmarshes. Mainly a non-breeding visitor; widespread but local, present Sept–Mar.

5. NORTHERN PINTAIL *Anas acuta* 51–66cm

Recognised by its long-bodied appearance. Distinctive male has chocolate-brown head and nape with white breast extending as a stripe up the side of the head. Plumage otherwise finely marked grey but shows cream and black at the stern and a long, pointed tail, often held at an angle. Female has mottled brown plumage. Male utters a whistling call. Favours well-vegetated wetlands. Widespread and locally common non-breeding visitor, present Oct–Mar.

6. MARBLED DUCK *Marmaronetta angustirostris* 40–42cm

Subtly marked but distinctive duck. Sexes are similar, the grey-brown ground colour marked with pale buff spots. Bill is dark. Shows a dark smudge through eye and an indistinct pale patch at the base of the bill. In flight, wings look uniformly brown. Favours well-vegetated wetlands and keeps close to cover. Rare resident and partial migrant in S Spain, NW Africa, Turkey and Middle East; some disperal and occasional breeding occurs elsewhere within these general areas.

1. COMMON POCHARD *Aythya ferina* 42–49cm

Bulky diving duck. Male has a reddish-orange head, a black breast, grey flanks and back, and a black stern. Female has a brown head and breast, grey-brown back and flanks, and a pale 'spectacle' around eye. Bill of both sexes is dark with a grey band near the tip. Favours lakes and reservoirs. Mostly silent. Widespread and locally common non-breeding visitor, present Oct–Mar; also a local breeding resident in small numbers.

2. FERRUGINOUS DUCK *Aythya nyroca* 38–42cm

Attractive diving duck. Male has chocolate-brown plumage that is darkest on the back and richest chestnut on the head; white stern is diagnostic. Bill is grey with a dark tip; eye has a white iris. Female is similar to male but plumage colours are duller and the eye is dark. In flight, shows striking white wingbar and white underwing. Favours well-vegetated lakes. Mainly silent in the region. Local resident along N coast of Mediterranean. More common and widespread as a non-breeding visitor, Sept–Mar (influx of birds from Central Europe).

3. TUFTED DUCK *Aythya fuligula* 40–47cm

Striking, yellow-eyed diving duck. Male looks black and white with a diagnostic tufted crest; head and neck show a purple sheen in good light. Female is brown, palest on the flanks; often shows a small crest and an indistinct patch of white at the base of the bill. Bill is grey and black-tipped in both sexes. In flight, white flight feathers with black trailing edge create a striking pattern in both sexes. Favours lakes and reservoirs and usually found in small flocks. Mostly silent. Best known as a widespread non-breeding visitor, present Sept–Mar (small numbers are resident in W, within species' winter range).

4. GREATER SCAUP *Aythya marila* 40–50cm

Robust yellow-eyed diving duck, both sexes of which recall their Tufted Duck counterparts. Male has a rounded, glossy green head (often looks black in poor light), black breast, white belly and flanks, grey back and black stern. Female is brown, palest on the flanks and with a striking white patch at the base of the bill. Favours coastal seas. Silent. Scarce winter visitor, present Oct–Mar, mainly to NE of region (Adriatic and Black Seas).

5. RED-CRESTED POCHARD *Netta rufina* 53–57cm

Sizeable and distinctive duck. Male has a colourful orange head and neck, and a striking red bill; body feathers are black except for grey-brown back and white flanks. Female has a pink-tipped dark bill, and a dark cap and nape; cheeks and throat are white while body plumage is brown. Wing markings are striking in flight: upperwings show white flight feathers with black trailing edge while underwings are pure white. Favours wetlands, often upending to feed on plants. Mostly silent. Local resident, mainly S Spain, S France and Turkey; occasional breeder elsewhere but more widespread in winter.

6. WHITE-HEADED DUCK *Oxyura leucocephala* 43–48cm

Distinctive 'stifftail' duck. Male has a white head, a black cap and eye, and a large, bright blue bill, swollen at the base; body plumage is brown, darkest on the neck. Female's bill is similar in shape to male's but dark grey; body plumage is brown while face is white with more extensive dark cap than on male and a dark eyestripe. Favours well-vegetated lakes. Mostly silent. Rare local resident and migrant visitor, mainly S Spain, NW Africa, Turkey; occasional breeder elsewhere. Some dispersal occurs in winter, particularly in E.

1. RED-BREASTED MERGANSER *Mergus serrator* 52–58cm

Grebe-like diving duck with a shaggy crest. Male has narrow red bill, green head, white neck and orange-red breast; shows grey flanks and a black back. Female has red bill and dirty orange head and nape, except for pale throat; body plumage is grey-buff. In flight, both sexes show white on inner wing. Favours coastal waters. Silent in the region. Local winter visitor, especially to Adriatic and Black Seas, present Oct–Mar.

2. GOOSANDER *Mergus merganser* 60–65cm

Elegant and comparatively large diving duck. Male has a glossy green head, white body, black back and bright red, serrated-edged bill. Appears black and white at a distance but, at close range, white underparts are seen to be flushed pink. Female has a similar reddish bill to the male but the head is orange-red with a shaggy crest. Body plumage is greyish. Both sexes show white on the inner wing in flight. Favours coastal lakes. Silent. Mainly a local winter visitor, present Oct–Mar; occasionally breeds within region.

3. COMMON GOLDENEYE *Bucephala clangula* 42–50cm

Attractive diving duck. Male has mainly black and white plumage with a large, rounded head; shows a yellow eye and a round white patch at the base of the bill. Female has grey-brown body plumage separated from the dark brown head by a pale neck; eye is pale yellow. Dives frequently and for long periods. Silent in the region. Favours lakes and coastal waters. Scarce winter visitor to N shores of Mediterranean and Black Sea, and to large freshwater lakes in the region; present Oct–Mar.

4. SMEW *Mergus albellus* 38–44cm

Small but striking diving duck. Male is unmistakable, looking pure white at a distance; close view reveals black patch around eye and black lines on back and breast. Female and immature birds show a reddish-orange cap that contrasts with the white cheek; body plumage is greyish. Favours fish-rich rivers and lakes; sometimes also on coastal waters. Regular winter visitor to region from Danube Delta south along Black Sea coast to Turkey and present Nov–Mar. Occurs elsewhere in the region irregularly and only moderately widespread during particularly harsh winters.

5. VELVET SCOTER *Melanitta fusca* 52–58cm

Attractive, mainly dark diving duck. Male has mostly black plumage, but the white eye, white patch under the eye and yellow colour on the bill are all visible at a considerable distance. Female is dark brown but shows a small but discrete pale patch at the base of the bill and on the cheeks. Both sexes have a striking white patch on the inner wing, this feature most obvious in flight but sometimes seen on swimming birds as well. Dives well and frequently. Silent in the region. Favours coastal waters. Rare winter visitor, mainly to N Mediterranean and Black Sea; present Oct–Mar.

6. COMMON SCOTER *Melanitta nigra* 45–55cm

Dumpy diving duck. Male is the only all-dark duck found in the region; black colour relieved by a yellow ridge on the bill, visible only at close range. Female has mainly dark brown plumage but cheeks appear paler grey-brown. Dives well and frequently. Silent in the region. Favours coastal waters. Scarce winter visitor to W Mediterranean, present Oct–Mar. More regular on Atlantic coasts.

1. LAMMERGEIER *Gypaetus barbatus* WS 265–280cm

A huge and distinctive vulture. Adult has proportionately long, narrow wings and a long, wedge-shaped tail. Seen from above, plumage appears mainly dark except for the buffish-orange head; from below, plumage can look all-dark but in good light the orange-buff head and underside of the body contrast with the dark wings and tail. A close view of an adult reveals a dark patch around the eye and dark, moustache-like feathers. The all-dark juvenile can be recognised by its flight silhouette alone. Soars effortlessly over cliffs and mountain tops. Feeds on carrion and specialises in dropping bones from a great height, causing them to shatter and hence permit the easy extraction of marrow. Rare resident of the Pyrenees, Corsica, Crete, N Greece, Turkey, NW Africa and Middle East; immatures sometimes wander beyond species' usual range.

2. EGYPTIAN VULTURE *Neophron percnopterus* WS 155–180cm

Europe's smallest vulture but still a sizeable bird of prey. Seen from below in flight, adult has dirty white plumage with black flight feathers; tail is white and wedge-shaped. From above, looks whitish with contrasting dark flight feathers. A close view of a standing adult bird reveals a bald yellow face and pinkish-yellow legs. The all-dark juvenile has a flight silhouette similar to that of the adult (it could conceivably be confused with a juvenile Lammergeier, but note that species' much larger size and its proportionately longer, narrower wings and tail). The white elements of an adult Egyptian vulture's plumage are acquired during successive moults over a five-year period. Favours mountains and gorges in warm, dry areas. Soars effortlessly. Summer visitor, present mainly Apr–Sept. Most numerous in Iberia but also occurs locally in N Africa and across Europe to Middle East.

3. EURASIAN BLACK VULTURE *Aegypius monachus* WS 250–295cm

A huge raptor and the largest vulture in the region. Has long, broad wings that are parallel-sided and square-ended; the primaries form distinct, splayed 'fingers'. Compared to the size of the wings, the head looks proportionately small and the tail relatively short. In flight, appears all dark although the plumage is in fact dark brown; pale lines are sometimes visible on the underwing coverts. A close view of a standing adult reveals a bald head and upper neck that are blackish below but pale above; the base of the bill is yellowish and the collar ruff is pale brown. Juvenile is similar to the adult but the underwing coverts are uniformly dark, the head, neck and collar ruff look uniformly dark, and the base of the bill is pinkish. Soars effortlessly at immense heights over mountains; wings usually held flat or slightly bowed downwards. Rare resident confined to S Iberia, Majorca (where it is perhaps easiest to observe), N Greece and Turkey.

4. EURASIAN GRIFFON VULTURE *Gyps fulvus* WS 240–280cm

A huge raptor with buffish-brown body plumage and wing coverts, and dark flight feathers. A close view of an adult reveals a bald, pale head and neck and a collar ruff of whitish-grey feathers; the bill is yellow. Juvenile is similar to the adult but the bill is grey and the collar ruff is buffish brown. In flight, all birds show wings that are long and broad at the base, but that taper to finger-like primaries at the tip; compared to the wing area, the head looks tiny and the tail appears short. Typically soars with wings in a shallow 'V'. Favours warm mountainous terrain. Resident in Iberia, Sardinia, N Greece, Turkey, Crete, Cyprus, the Middle East and NW Africa; immatures sometimes wander beyond typical range.

1. NORTHERN MARSH HARRIER *Circus aeruginosus*
WS 115–130cm

The most regularly encountered harrier across much of the region. Adult male is reddish brown except for blue-grey head and grey unbarred tail; in flight, has grey and brown patches on wings, and black wingtips. Adult female and immature birds (of both sexes) are dark brown except for pale leading edge to wing and pale cap and chin (tail dark brown in immature, reddish in adult female). Typically flies low over marshes, but equally at home over grassland and arable fields. Common passage migrant throughout the region and occurs year-round in some areas, notably Iberia, NW Africa and Balkan region; locally common as a breeding species throughout and widespread during the winter months, especially in the E.

2. PALLID HARRIER *Circus macrourus* WS 100–115cm

Graceful bird of prey and generally the hardest of the harriers to observe in the region. Adult male is mainly pale grey (often so pale as to appear almost white) with black wingtips and a white rump. Note that the dark wingtips form a narrow wedge shape, much less extensive than the black wingtips of a male Hen Harrier; the white rump is proportionately smaller and more difficult to see than in male Hen Harrier because of Pallid's overall paler body colour. Adult female is very similar to its Hen and Montagu's harrier counterparts. Note that the face looks owl-like owing to the pale rim to the facial disc; also, from below and in flight, the secondaries are dusky and hence look less strikingly barred. Juvenile resembles juvenile Montagu's, with orange-red body and underwing coverts; close view reveals the darkish head to be defined by a pale collar. Favours grassland and arable fields. Scarce passage migrant (Mar–May and Aug–Oct), and scarce winter visitor, mainly in E of region.

3. HEN HARRIER *Circus cyaneus* WS 100–120cm

Elegant adult male is essentially pale grey with black wingtips and a white rump. Superficially similar to adult male Montagu's and Pallid harriers, but lacks the former's dark wingbars and red streaking and is much darker grey overall than the latter species. Adult female is brown, streaked on underside and with a barred tail and white rump. Juvenile is similar to adult female although colour of body and underwing coverts is usually more reddish brown. Favours grassland, marshes and arable fields, quartering the ground in leisurely flight. Occurs year-round in parts of Iberia and S France but best known as a widespread but rather scarce winter visitor and passage migrant (mainly Oct–Apr).

4. MONTAGU'S HARRIER *Circus pygargus* WS 105–120cm

Graceful and elegant bird of prey. Adult male is mainly blue-grey. In flight and from below note the red-flecked underparts, red-barred wing coverts, barred tail, dark wingtips and two black wingbars across the secondaries. From above, note the dark wingtips, the single black wingbar across the secondaries and the white rump; note also the contrast between the darker grey innerwing coverts and back, and the paler secondaries and primary coverts. Adult female is brown with a barred tail, streaked underparts, barred underwing coverts and a white rump. Juveniles (of both sexes) resemble adult female although underparts (except flight feathers and tail) are overall more reddish brown and unstreaked. Immature male acquires full adult plumage in successive moults over two years. Favours open grassy places, including arable fields, flying at a leisurely pace low over the vegetation, stalling and dropping to catch small mammals, birds and insects. Summer visitor to the region (common as a breeding species only in Iberia) and a widespread passage migrant; present Apr–Sept.

1

1

imm

♂

♀

2

♂

2

♀

2

3

3

3

3

♂

3

♀

♂

♀

imm 4

4

1. BLACK KITE *Milvus migrans* WS 145–165cm

Medium-sized raptor. Recalls female Marsh Harrier. Adult is dark grey-brown, palest on head; has a black-tipped yellow bill and yellow legs. Juvenile is paler than adult. In all birds, tail is forked (can look straight-ended when fanned). Wings are usually held flat when soaring, and twists fanned tail. Favours wooded wetlands but will scavenge at rubbish dumps. Best known as a widespread passage migrant and summer visitor; present Apr–Sept.

2. RED KITE *Milvus milvus* WS 155–175cm

Graceful raptor. Resembles Black Kite but more colourful and with more deeply forked tail. Adult is mainly reddish brown with grey head, broad pale patch near wingtip and pale undertail. From above, tail is red and flight feathers are dark. Juvenile is similar to adult but reddish elements to plumage are subdued. Soars effortlessly and twists tail. Call resembles a person whistling for their dog. Favours wooded valleys for nesting and open country for feeding. Resident, mainly from Iberia to S Italy; local in NW Africa and elsewhere. Disperses in winter and migrants from N appear (sometimes more widespread than map suggests).

3. BLACK-WINGED KITE *Elanus caeruleus* WS 70–85cm

Elegant bird of prey. Adult has blue-grey, white and black plumage. In flight and from above looks grey except for black forewing and white outer tail; from below, looks extremely pale with black wingtips. Close view reveals yellow-orange legs, red eye with black 'eyebrow', and yellow base to bill. Juvenile is similar to adult but back is 'scaly' and shows buff on crown and breast; in flight from below, all flight feathers are dark. Favours grassy areas with scattered trees. Local resident in Iberia (range may be expanding), N Africa and the Middle East.

4. NORTHERN GOSHAWK *Accipiter gentilis* WS 100–115cm

Similar to Eurasian Sparrowhawk but larger (female is almost buzzard-sized) and with relatively bulkier body, shorter tail and longer wings. Adult has grey-brown upperparts and pale, dark-barred underparts. Shows staring yellow eyes, white stripe over eyes and yellow legs. Often soars, revealing fluffy white undertail feathers. Favours extensive forests. Widespread but scarce resident from Iberia to S Turkey; very local elsewhere.

5. LEVANT SPARROWHAWK *Accipiter brevipes* WS 65–75cm

Recalls Eurasian Sparrowhawk. Adult male has barred reddish underparts, and blue-grey upperparts including crown and cheeks; in flight, underwings look pale with striking dark tips. Adult female has barred reddish-brown underparts and grey-brown upperparts; note dark vertical line on pale throat. Juvenile mostly brown; pale underparts have dark streaks and shows a dark throat line. All birds have dark eyes. Favours wooded areas. Migrates in flocks. Summer visitor (May–Aug), range centred on Greece and Bulgaria. Passage migrant in Middle East.

6. EURASIAN SPARROWHAWK *Accipiter nisus* WS 60–75cm

Dashing raptor with short, rounded wings and long, barred tail. Male is smaller than female and has blue-grey upperparts; underparts (including wings) are barred reddish brown. Female has grey-brown upperparts and pale underparts with brown barring in all areas. Juvenile is brownish with strongly barred underparts. All birds have yellow to yellow-orange eyes. Hunts in low-level flight for small birds. Soars on migration and in territorial display. Favours woodland and scrub. Widespread resident, numbers boosted in winter by migrants from N Europe.

1

2

3. juv

3. juv

3

5

5

4

5

6

6

1. LONG-LEGGED BUZZARD *Buteo rufinus* WS 130–155cm

Similar to pale forms of Common Buzzard and Steppe Buzzard but longer winged. From below, body and wing coverts of adult are unstreaked reddish brown, darkest on belly; tail uniformly pale reddish and unbarred, and head often looks pale. Flight feathers whitish except for contrasting black wingtips and dark trailing wing edge; note also the dark carpal patch. Juvenile similar to adult but generally much paler. Favours dry rocky terrain and mountains. Local summer visitor, mainly Greece to Romania, present May–Sept. Resident from Turkey eastwards and in N Africa. Passage migrant throughout the Middle East.

2. COMMON BUZZARD *Buteo buteo* WS 115–125cm

Variable medium-sized raptor with broad wings and a short tail, fanned when soaring. Typical adult has pale underwings except for the dark trailing edge and coverts; body usually dark brown except for pale band on chest. Appears uniformly brown from above. Plumage varies, however, and some birds are mostly white while others are uniformly dark. So-called **Steppe Buzzard** *B. b. vulpinus* resembles Long-legged Buzzard, with rufous body, underwing coverts and tail; note, however, dark terminal band on tail. Juveniles of all races are less well marked than their adult counterparts and dark terminal tail band is absent. All birds soar with wings in a 'V' shape. Utters mewing calls. Favours wooded areas. Widespread resident. Steppe Buzzard is a passage migrant through the Middle East.

3. EUROPEAN HONEY BUZZARD *Pernis apivorus*
WS 135–150cm

Medium-sized bird of prey. Resembles Common Buzzard but pale underparts show more contrasting dark barring and a darker carpal patch on the wings. Tail is proportionally long with two broad dark bands at the base and a dark terminal band. In flight, head looks narrow and pale. Soars on flat wings. Favours vast forests; feeds mainly on the larvae of wasps and bees, which are excavated from subterranean nests. Summer visitor to N of region, mainly from N Iberia to Greece; present mid-Apr–Sept. Widespread passage migrant throughout the region.

4. GOLDEN EAGLE *Aquila chrysaetos* WS 190–230cm

Large raptor with proportionally long tail and wings, the latter narrowing at the base. Adult has mainly dark brown plumage with paler feathers on back, and a golden-brown crown and nape. In flight, looks dark from below with the reddish-brown head, breast and wing coverts sometimes discernible; from above, note pale panel on wing coverts and pale base to tail. Juvenile looks dark except for striking white wing patches, and a tail that is white with a black terminal band. Adult plumage is acquired in successive moults over four to five years. Soars on parallel-sided wings held in shallow 'V'. Local resident in mountains.

5. EASTERN IMPERIAL EAGLE *Aquila heliaca*
WS 180–215cm

Adult is dark brown with white shoulder patch and pale buffish crown and nape. In flight, from below shows broad, parallel-sided wings and pale grey, finely barred tail with dark terminal band; from above, note similar tail pattern but also white shoulder patch. Juvenile has pale buffish body and wing coverts, contrasting with dark flight feathers and tail; from above, note pale rump. Favours plains and wooded slopes. Scarce resident in E; passage migrant and winter visitor to Middle East. **5a. Spanish Imperial Eagle** *A. (h.) adalberti* is now treated as a separate species. Adult recalls Eastern Imperial but note the pale leading edge to wing seen from below and above in flight. Juvenile rufous with darker flight feathers. Rare resident in S Iberia.

2. Common

2. Steppe

1

2

4

3

4

juv

ad

5a. juv

5a. ad

5

1. TAWNY EAGLE *Aquila rapax* WS 165–185cm

Impressive bird of prey. Wings are rather broad and often appear square-ended and parallel-sided when soaring, with distinct 'fingers' formed by primary projections. Tail is proportionately short and often fanned out when soaring. Adult has head, body and wing coverts that are tawny brown; the colour contrasts with the dark flight feathers and tail. Seen from above and in flight, note the pale rump. Feet and base of bill are yellow; at close range, pale eye can be seen. Juvenile is paler on body and wing coverts than adult and so, from below, the contrast with dark flight feathers and tail is more striking; from above and in flight, note white wingbar and trailing edge to the wings. Favours dry, lightly wooded slopes. Local resident in NW Africa.

2. STEPPE EAGLE *Aquila nipalensis* WS 165–200cm

Similar to Tawny Eagle but larger, with proportionately longer wings, and generally darker; also widely separated geographically in the region. Adult plumage is generally dark brown. In flight and from below, can look uniformly dark although marginally darker carpal patches and breast are sometimes discernible; from above, often looks uniformly dark although pale rump and pale base to primaries are sometimes seen. Feet and base of bill are yellow. Dark eye discernible at close range. Juvenile has reddish-brown body plumage and wing coverts. In flight, seen from below, wing coverts are separated from dark flight feathers by a striking white band, and wings also have a white trailing edge; from above, shows white trailing edge to wings, and white at base of primaries continuing as white band separating wing coverts from secondaries. Favours open, dry country. Common passage migrant in the Middle East, mainly Mar–Apr and Sept–Nov.

3. GREAT SPOTTED EAGLE *Aquila clanga* WS 155–175cm

Large bird of prey. Wings are proportionately broad and parallel-sided when soaring; appear rather broad and square-ended but splayed 'fingers' of primary projections are easily seen. Adult is mainly dark brown, with yellow legs and yellow base to bill. In flight, from below, wings look uniformly dark or with flight feathers marginally paler than wing coverts; pale patch at base of primaries is sometimes visible. From above, adult looks mainly dark except for pale base to tail and faint pale bases to primaries. Juvenile is dark brown, marked with white teardrop spots on back and upperwing coverts. In flight and from below, shows marked contrast between dark wing coverts and body, and paler flight feathers; from above, shows white bars on wing coverts and back, and white rump and trailing edges to wing and tail. Favours wetland forests. Passage migrant in E of region; also a local and scarce winter visitor, mainly Oct–Mar.

4. LESSER SPOTTED EAGLE *Aquila pomarina* WS 145–165cm

Similar to Great Spotted Eagle; subtle structural differences include proportionately narrower wingtips, less evident 'finger' projections of primaries and less robust bill of Lesser Spotted. In flight and from below, adult has flight feathers and tail that look darker than body and wing coverts; two parallel pale lines at base of primaries are sometimes discernible. From above, adult looks dark except for pale rump and patch at base of primaries. Juvenile is dark but with white teardrop spots on back and upperwing coverts. In flight and from below, looks similar to adult but with pale trailing margin to wings and tail; from above, shows white bar at base of flight feathers, pale rump and white-tipped tail. Favours open country and woodland. Summer visitor and passage migrant in E of region, present Apr–Sept.

1

2

juv 3

3 ad

ad

juv

ad

juv

juv 4

4 ad

1. BONELLI'S EAGLE *Hieraaetus fasciatus* WS 150–170cm

Large and impressive raptor. Adult has brown upperparts, with white patch on upper mantle, and pale underparts marked with dark streaks. In flight, from below, note pale belly and contrast between the black wingbar and paler flight feathers and leading edge to inner wing; tail is grey, with dark terminal band. From above, appears mainly dark except for pale patch on back and paler tail with dark terminal band. In flight, from below, most juveniles have a rufous body and wing coverts (white in some juveniles), while the flight feathers and tail are barred grey. From above, all juveniles are uniform brown. Favours wooded mountains. Widespread but local resident, most numerous in Iberia.

2. BOOTED EAGLE *Hieraaetus pennatus* WS 100–130cm

Superficially buzzard-like raptor but with longer, splayed primaries and proportionately longer tail. Occurs in two colour forms. Pale-phase adult has faintly streaked whitish underparts and buffish-brown upperparts; in flight, from below, the dark flight feathers contrast markedly with the pale body and wing coverts. Dark-phase adult is uniformly dark brown. Both colour phases show a pale 'V' on the back and upperwings. Juveniles of both phases are similar to their respective adults but colours and markings are more subdued. Favours wooded hills. Summer visitor to the region, present Apr–Sept. Widespread but perhaps commonest in Iberia, Majorca, Greece and Turkey.

3. SHORT-TOED EAGLE *Circaetus gallicus* WS 170–185cm

Large and distinctive raptor. Wings are long, broad and pale and tail is relatively long but broad. Close view of perched bird reveals relatively large head (almost owl-shaped). Adult has brown upperparts and dark-barred pale underparts. In flight, from below, looks rather uniformly pale and barred, except for dark head and neck; tail often shows a dark terminal band. Juvenile is similar to adult but paler overall. Often hovers, scanning ground for snakes (its favourite prey). Favours open, lightly wooded hillsides. Summer visitor, from Iberia to the Middle East and NW Africa; present Apr–Sept.

4. WHITE-TAILED EAGLE *Haliaeetus albicilla* WS 190–235cm

Immense bird of prey with broad and proportionately long, parallel-sided wings and conspicuous primary projection 'fingers'. Adult has buffish-brown plumage (palest on the head) except for the pure white, wedge-shaped tail. Head and yellow bill look proportionately large. Juvenile has darker body plumage than adult and dark wedge-shaped tail; bill is pale-based but otherwise grey. Adult plumage acquired in successive moults over four to five years. Favours extensive wetlands. Active predator of fish and waterbirds but will also scavenge. Local resident in E, notably N Greece, Turkey and Danube Delta. Numbers boosted, and range expanded, in winter months by visitors from central Europe.

5. OSPREY *Pandion haliaetus* WS 145–160cm

Distinctive fish-eating raptor. Perched adult shows dark brown back and wings, whitish underparts, and pale head except for dark stripe through eye. In flight and from below, note the long, narrow wings, and the overall pale appearance except for the dark primaries and dark carpal patch. Upperparts are brown except for the pale crown. Juvenile is similar to adult but upperparts appear scaly and underwings are less strikingly marked than on adult. Associated with fish-rich water bodies. Frequently hovers and plunge-dives, feet first, for prey. Widespread passage migrant; small numbers are resident or winter in W of region.

2. pale phase

2. dark phase

juv

1

2

4. ad

3

4. juv

4

5

5

1. LESSER KESTREL *Falco naumanni* WS 58–72cm

Small, elegantly proportioned falcon. Superficially similar to Common Kestrel in all plumages. Adult male has unspotted chestnut back, blue-grey inner wing and head, dark outer wing, and pale grey tail with dark terminal band; upperwing pattern useful for distinction from male Common Kestrel. From below, adult male has almost whitish underwings with contrasting black wingtips. Adult female and juveniles (of both sexes) have brown plumage marked with dark spots, except for dark primaries and barred tail; similar to, but paler overall than, female Common Kestrel. Usually seen in flocks and breeds colonially in old buildings or on cliffs. Hovers, but much less frequently than Common Kestrel. More vocal than Common Kestrel, typical call a rasping *tchee-tchee-tchee*. Mainly a passage migrant and summer visitor, present Apr–Sept; however, small numbers winter in S Spain and N Africa. Locally common at strongholds in Iberia, Greece and Turkey but otherwise generally scarce.

2. COMMON KESTREL *Falco tinnunculus* WS 65–80cm

The most widespread and familiar small raptor in the region. Adult male has a spotted, orange-brown back, a blue-grey head and a blue-grey tail with a terminal black band. In flight, from above, wing pattern comprises the orange-brown back and inner wing, and the dark outer wing (cf male Lesser Kestrel); from below, wings appear uniformly greyish and barred. Adult female and juveniles (of both sexes) have brown upperparts and barred, pale grey-brown underparts. Nests in trees and on cliff ledges, but also in man-made settings such as loft spaces in barns. Feeds primarily on small mammals but takes insects in summer months. Generally solitary. Frequently hovers using headwind or updraught for assistance. Typical call a shrill *kee-kee-kee*. Common resident and passage migrant throughout the region, including most islands.

3. RED-FOOTED FALCON *Falco vespertinus* WS 65–75cm

Small and elegant falcon that looks elongated and long-winged when perched. At a distance, and in flight silhouette, can be mistaken for either Common Kestrel or Eurasian Hobby. Adult male has mainly dark grey plumage with pale primaries, red vent and thighs, and red feet; red skin around eye is visible at close range. Immature male is similar but has paler underparts and pale face and throat. Adult female has orange-buff crown and underparts (including wing coverts but excluding flight feathers and tail, which are barred grey); back is barred grey and has dark mask through eye. Juvenile resembles adult female but upperparts are grey-brown and underparts are pale and heavily streaked. Migrant birds typically favour open fields and marshes. Perches on wires but also hovers. Breeds colonially and usually seen in flocks on migration. Typical call a rapid *kee-kee-kee*. Common passage migrant in E half of region; also seen W to Iberia in smaller numbers, particularly in spring.

4. MERLIN *Falco columbarius* WS 55–70cm

Small, dashing raptor and the smallest bird of prey in the region. Adult male has blue-grey upperparts and buffish underparts that are streaked and spotted. Adult female and juveniles (of both sexes) have brown upperparts and pale underparts marked with large brown spots. Typically seen flying low over ground in dashing flight, pursuing prey such as pipits and small finches. Seldom soars and often perches on fenceposts and rocks for long periods. Favours open country and coastal plains and marshes. Generally silent in the region. Winter visitor to the region and widespread but local throughout; present Oct–Mar.

subad ♂

1. EURASIAN HOBBY *Falco subbuteo* WS 70–85cm

Elegant falcon. Adult is dark blue-grey above and pale, dark-streaked below. At close range, note dark 'moustache', white cheeks and reddish 'trousers'. Juvenile is similar to adult but colours are subdued; lacks red 'trousers'. In flight, all birds have anchor-like silhouette with narrow, swept-back wings and relatively long tail. Hunts insects and birds. Favours open country with scattered trees. Widespread summer visitor and passage migrant, present May–Sept.

2. PEREGRINE *Falco peregrinus* WS 95–115cm

Robust and impressive falcon. Adult has dark blue-grey upperparts and pale, barred underparts; face has dark crown and mask. Juvenile is browner overall than adult, with streaked underparts. Favours mountains, sea cliffs and coastal marshes. Soars on bowed wings but dives with wings swept back. Local resident from Iberia and N Africa to Turkey. More widespread in winter. The **Barbary Falcon** *F. (p.) pelegrinoides* (not illustrated) recalls both Peregrine and Lanner. Has blue-grey upperparts and pale buff, finely barred underparts; shows rufous on sides of nape and dark grey crown (rufous in some Lanner races). Favours deserts and foothills. Resident in NW Africa and the Middle East.

3. SOOTY FALCON *Falco concolor* WS 80–90cm

Elegant falcon. Flight silhouette recalls Eleonora's Falcon but tail is shorter. Adult has blue-grey plumage; close view reveals yellow legs, eye-ring and cere. Juvenile has scaly grey upperparts, streaked buffish underparts and dark facial mask; note tail's dark terminal band, seen from below in flight. Favours cliffs and mountains in deserts and on coasts. Summer visitor to S Israel and NE Egypt, present May–Oct.

4. ELEONORA'S FALCON *Falco eleonorae* WS 90–105cm

Resembles Eurasian Hobby but has proportionately longer tail and wings. Adults occur as two colour forms. Dark-phase birds are uniform dark brown. Pale-phase birds show pale cheek, dark 'moustache' and pale underside with dark streaks; plumage is otherwise dark. Juvenile resembles pale-phase adult but colours and markings are subdued. Catches migrant songbirds, although insects are also taken in spring. Breeds on sea cliffs but often hunts inland prior to nesting. Present May–Oct, with scattered colonies from Majorca and N African coast to Cyprus. Nests in late summer.

5. SAKER FALCON *Falco cherrug* WS 105–125cm

Large, broad-winged falcon. Adult is brown above and dark-streaked buffish below; head (including crown) is pale. In flight and from above, reddish-brown back and wing coverts contrast with dark flight feathers; from below, darkish body and wing coverts contrast with paler flight feathers and tail. Juvenile is darker overall than adult, with heavily streaked underparts and darker facial markings and crown. Favours open country with scattered trees. Mainly a winter visitor and passage migrant to E but also a scarce and local resident breeder.

6. LANNER FALCON *Falco biarmicus* WS 95–115cm

Similar to Peregrine but slimmer and with longer tail; wings rather broad with rounded tips. In flight, European adults are dark grey above with dark crown and 'moustache', and orange-buff nape; N African birds are blue-grey above with orange-buff crown and dark moustache; all adults are pale below with dark spots. Juvenile is dark brown above and heavily streaked below. Favours arid, mountainous terrain and steppe habitat. Local resident.

1

2

3

4

4

pale phase

dark phase

5

juv 5

6

1. DOUBLE-SPURRED FRANCOLIN *Francolinus bicalcaratus* 30–32cm

Robust, shy gamebird. Adult appears brown but close view reveals subtle and beautiful black markings on body feathers; these form dark streaking on the underparts. Neck and crown are reddish brown and shows pale supercilium and throat. Tail and lower rump are uniform brown. Sexes are similar but male has spurs on legs. Favours farmland and open woodland. Shy and difficult to see. Male utters rasping and repetitive *koo-arr koo-arr* call. Rare and local resident in Morocco.

2. BLACK FRANCOLIN *Francolinus francolinus* 33–36cm

Distinctive gamebird. Shy and more often heard than seen: male utters grating call from cover in spring. Adult male is mainly black except for white cheeks, chestnut collar and brown wings; body is spangled with white spots. Female is mainly brown, the body feathers boldly marked with black and white; head is paler than body and shows a chestnut collar. In flight, all birds reveal dark outer tail feathers and brown wings. Favours arable land and scrub. Resident on Cyprus and, locally, from SE Turkey eastwards; introduced to Tuscany.

3. ROCK PARTRIDGE *Alectoris graeca* 33–35cm

Dumpy gamebird that shares the characteristic red bill, eye-ring and legs with other *Alectoris* partridges. Note the white throat, defined by the neat black border that runs through the eye to the bill; base of bill is entirely bordered by black. Body plumage is blue-grey, grading to buffish on the belly and with striking black and white markings on the flanks. Males utters rasping, repetitive *che-cheer-it-chee* calls. Favours rocky mountain slopes, usually at moderate to high altitudes. Resident in suitable habitats within range.

4. RED-LEGGED PARTRIDGE *Alectoris rufa* 32–34cm

Plump gamebird with red bill, eye-ring and legs. White throat is bordered with gorget of black spots; plumage is otherwise mainly blue-grey and warm buff except for black and white barring on flanks. Favours arable land, marshes and foothills. Male utters repetitive call *ca-chah-chah, ca-chah-chah*. Seen in small parties outside breeding season. Widespread resident within range. Other potentially confusing *Alectoris* species can be separated by careful study of plumage, but also by habitat preference and geographical range.

5. BARBARY PARTRIDGE *Alectoris barbara* 32–35cm

Distinctive *Alectoris* partridge with typical red bill, eye-ring and legs. Note, however, the pale blue-grey throat and supercilium, the pale brown eyestripe and the border to the throat that is broad, brown and marked with white spots. Body plumage is blue-grey and buff; markings on flanks appear brownish, against a pale buff background. Favours vegetated lowlands to lower mountain slopes. Male utters rasping *ka-cheh-cheh* calls. Widespread resident in N Africa; also found on Gibraltar and Sardinia.

6. CHUKAR *Alectoris chukar* 32–35cm

Similar to Rock Partridge but note the creamy (not white) throat, the white (not black) border to the base of the upper mandible, and the broad, pale supercilium. Body plumage is otherwise blue-grey, grading to buff on belly and with typical *Alectoris* black and white markings on flanks. Male typically utters repetitive *chuk-ar, chuk-ar* calls. Favours warm hillsides (sometimes on rocky lower slopes of mountains) and agricultural land. Resident. Natural distribution (shown on map) is confused by introductions elsewhere (not shown).

1. GREY PARTRIDGE *Perdix perdix* 29–31cm

Compact gamebird. Adult male has mainly blue-grey plumage with fine vermiculation markings visible at close range. Note also the orange-buff face, the dark maroon-chestnut mark on the belly, maroon stripes on the flanks and the brown-streaked back. Female is similar to male but markings are less distinct. Juvenile is brownish buff, streaked on the flanks and with scaly appearance to the back. Often seen in small groups. When disturbed, takes to the air on whirring, noisy wings. Song is a rasping, strangled *kee-er-ick*. Favours arable fields and grassland. Locally common resident.

2. SAND PARTRIDGE *Ammoperdix heyi* 22–25cm

Plump-bodied gamebird. Male has mainly sandy-brown body plumage but the flanks are marked with concentric wavy lines of white, brown and black. Head and neck are uniform blue-grey with white patches on cheeks and lores. Bill is yellow and legs are dull yellow-buff. Shows reddish outer tail feathers in flight. Female plumage recalls that of male but colours and flank patterns are more subdued; head and neck are uniform pale blue-grey except for whitish supercilium. Favours arid terrain. Resident in the Middle East. **See-see Partridge** *A. griseogularis* (not illustrated) occurs in similar terrain from SE Turkey eastwards. Male resembles male Sand Partridge but neck is marked with white spots, and face has black forehead and mask with white patch behind the eye.

3. COMMON PHEASANT *Phasianus colchicus* 53–59cm

Familiar gamebird. Unmistakable adult male has orange-brown body and long, orange tail. Head has blue-green sheen and red fleshy wattle; some birds have a white collar. Adult female is mottled buffish brown with a shorter tail than male. Juvenile resembles short-tailed female. When disturbed, flies off explosively on whirring wings, uttering loud shrieking call. Favours wooded farmland and scrub. A long-established introduction, now resident from Iberia and France eastwards, although absent from much of SE Europe.

4. COMMON QUAIL *Coturnix coturnix* 16–18cm

Tiny, short-tailed gamebird. More often heard than seen, except when flushed from cover. Plumage is essentially brown above and pale buff below. Male has streaked flanks and black, brown and white head markings including a black throat. Female is similar to male but markings are less striking and face shows pale throat. In flight, all birds show relatively long, rounded wings. Male utters repetitive *wet-my-lips* call on breeding territory. Favours arable fields and grassland. Widespread summer visitor and passage migrant, present Apr–Sept.

5. SMALL BUTTON-QUAIL *Turnix sylvatica* 15–16cm

Enigmatic and little-known bird; formerly called Andalusian Hemipode. Looks strikingly short-tailed and could be confused with Quail. Adult female has scaly-looking brown upperparts; underparts are paler buffish grey, flushed orange on the breast and with dark spots on the flanks. Eye is pale. Adult male is similar to female but colours and markings are more subdued. Favours open country with low, often grassy, vegetation. Secretive and hard to observe. If flushed (a rare event), typically alights standing bolt upright for a brief moment. Roles of sexes are reversed. Female utters cow-like mooing calls at dawn and dusk; an excellent ventriloquist making calls almost impossible to pinpoint. Presumed still to be resident in N Africa and possibly still occurs in SW Iberia although exact status is uncertain. Consider yourself privileged if you see one.

RAILS AND CRAKES

1. WATER RAIL *Rallus aquaticus* 23–28cm

Shy wetland bird. Pig-like squealing calls heard more often than bird itself is seen. Body is rounded in profile but flattened laterally, allowing passage through dense vegetation. Underparts are mainly blue-grey with black and white barring on flanks; upperparts are reddish brown. Has long, reddish bill and reddish legs with long toes. Favours reedbeds and marshes in summer and winter; on migration sometimes found in more open wetland habitats. Local resident throughout. More widespread in winter owing to influx of birds from N Europe.

2. CORN CRAKE *Crex crex* 22–25cm

A secretive bird that prefers to escape from danger by slipping away through vegetation. Upperparts are brown with black spots. Underparts and face are bluish grey with chestnut and white barring on flanks. Bill is stout and pink. Occasionally flushed from cover, when chestnut wings and long, trailing legs are noticeable. Favours meadows and other dry grassy areas. Mainly a passage migrant through the region but small numbers remain to breed, particularly along Adriatic coast, in S France and N Italy; present Apr–Sept. Male utters repetitive *crex-crex* call in breeding season, mainly after dark, but passage migrants are generally silent.

3. SPOTTED CRAKE *Porzana porzana* 20–24cm

Small, dumpy waterbird. Adult has mainly brown upperparts but feathers are spangled with white spots. Shows blue-grey on supercilium, throat and breast with black at base of bill. Sexes similar but plumage of male is brighter, and shows more contrast, than that of female. Juvenile is similar to adult but lacks blue-grey elements on face and breast. All birds have stubby yellow bill with red base, and greenish legs with long toes. Favours well-vegetated wetland margins. Male utters whiplash call in breeding season. Widespread passage migrant and local breeding species, present May–Sept. May winter locally in small numbers.

4. LITTLE CRAKE *Porzana parva* 18–20cm

A tiny waterbird, superficially similar to Baillon's Crake but marginally larger and body appears more elongated owing to longer tail and primary feathers. Adult male has mainly dark brown upperparts with pale and dark streaks; underparts including face are blue-grey and almost unmarked. Adult female has buffish-grey underparts and pale bluish grey on face; upperparts are brown and streaked. Juvenile is similar to female but underparts are paler with barring on flanks. All birds have disproportionately long toes and a yellowish bill that is red at the base. Favours marshes with dense cover. Territorial male utters soft yapping call. Mainly a summer visitor and passage migrant; commonest in E; present Apr–Sept. Scarce breeding species; occasional in winter.

5. BAILLON'S CRAKE *Porzana pusilla* 16–18cm

The smallest crake in the region. Adult resembles a miniature Water Rail with a much shorter bill. Adult has rich brown upperparts with dark and white streaks. Underparts, including face, are bluish grey with barring on flanks. Sexes are similar but female sometimes has paler throat than male. All birds have disproportionately long toes and a greenish bill that lacks red at base. Juvenile has streaked brown upperparts, underparts that are pale and strongly barred, and a dark-tipped bill. Male utters frog-like croaking call in breeding season. Favours waterlogged marshes with dense cover but sometimes seen in open wetlands on migration. Widespread but scarce passage migrant and summer visitor, present Apr–Sept. Very local breeding species; occasional in winter.

1. EURASIAN COOT *Fulica atra* 36–38cm

Has all-black plumage with white bill and frontal shield on head. Lobed toes aid swimming. Shows white trailing edge to inner wing in flight. Call is a loud *kwoot*. Feeds by upending or making shallow dives in water but also grazes on waterside grass. Favours lakes and marshes with open water during breeding season. At other times of year, also found on reservoirs, usually in sizeable flocks. Locally common resident; more widespread and numerous in winter.

2. RED-KNOBBED COOT *Fulica cristata* 38–42cm

Similar to Eurasian Coot but larger. Adult has red knobs over white facial shield; these are obscure at a distance and in winter when colour fades. In flight, lacks Coot's white trailing edge to secondaries. Found on well-vegetated lakes. Upends for food and dives well. Calls include a peculiar 'mooing' sound. Very local resident in S Spain and in NW Africa; seldom seen elsewhere.

3. PURPLE SWAMP-HEN *Porphyrio porphyrio* 45–50cm

Common Moorhen-like in outline but with striking colours. Adult has violet-blue plumage with blue face and red eyes, legs and bill. Undertail is white and exposed as bird nervously flicks its tail. Juvenile is grey with pale throat and dull red legs and bill. Calls include loud hoots and rasping rattles. Favours brackish pools with standing water, reedbeds and freshwater marshes. Rare and local resident. Reintroduced and easy to see at Albufera on Majorca.

4. COMMON MOORHEN *Gallinula chloropus* 32–35cm

Familiar waterbird. Adult is dark grey-black except for brownish wings and back. Has yellow-tipped red bill and frontal shield on head, white feathers on sides of undertail and white line along flanks. Legs and long toes are yellowish. Juvenile has pale brown plumage. Swims with jerky movement and flicking tail; legs dangle in flight. Call is a loud *kyrruk*. Widespread wetland resident; numbers boosted in winter by migrants.

5. LITTLE BUSTARD *Tetrax tetrax* 40–45cm

Adult is marbled grey-brown above and white below; male has black and white neck markings. In flight, all birds show mainly white wings with black wingtips. Favours open grassland and steppe habitat. Forms flocks outside breeding season. Locally common resident in Iberia. Scattered populations elsewhere, including NW Africa, S France and Sardinia.

6. GREAT BUSTARD *Otis tarda* 80–100cm

Huge and unmistakable. Male is larger than female and has marbled brown upperparts and white underparts. In breeding season, male displays with bulging neck and cocked-up tail; note white whiskers and blue-grey and rufous neck. Female and non-breeding males lack whiskers and have subdued neck colours. In flight, all birds show black and white wings. Favours open fields. Mostly silent. Wary. Local resident, mainly S Iberia and E Europe.

7. HOUBARA BUSTARD *Chlamydotis undulata* 55–65cm

Sizeable, wary bird. Upperparts are mainly barred brown while underparts grade from grey on neck to white on underparts. Adult has black line down side of neck. Wings show striking black and white pattern in flight. Favours semi-desert and steppe habitats. Local resident in N Africa and Middle East. (Middle Eastern race has recently acquired species status: Macqueen's Bustard *C. macqueenii*.)

ad with juv

1

2

4

4

4. juv

3

6

6

6

♂

6

♀

♂

5

5

7

1. BLACK-WINGED STILT *Himantopus himantopus* 35–40cm

Unmistakable black and white wader with extraordinarily long, bright red legs. Adult has black wings and mantle and white underparts; white head and neck show variable amounts of black (black is more extensive in male than female). Bill is long, needle-like and black. Juvenile is similar to adult but black elements of plumage show pale feather margins. In flight, all birds show a white rump and lower back, and long, trailing legs; wings are uniform black, both above and below, although in juveniles a pale trailing edge can be seen. Call is a shrill *kyipp*. Favours coastal lagoons and shallow lakes; often nests beside saltpans. Locally common summer visitor, present Mar–Oct; also overwinters in small numbers locally, eg S Iberia, NW Africa and Middle East.

2. PIED AVOCET *Recurvirostra avosetta* 43–45cm

Unmistakable pied wader. Adult has black-and-white plumage, long, blue legs and a long, upcurved bill that is swept from side to side through shallow water when feeding. Webbed toes aid walking on soft mud and swimming. Juvenile is similar to adult but dark elements of body plumage are dark brown and more diffuse. Call is a ringing *klu-ut, klu-ut*. Favours coastal lagoons, saltpans and estuaries. Local resident with scattered breeding colonies, mainly along European shores; more widespread in winter, when it forms flocks.

3. EURASIAN OYSTERCATCHER *Haematopus ostralegus* 41–43cm

Striking and robust pied wader. Summer adult has black head, neck and upperparts and white underparts. Bill is uniformly bright red and adult has red eyes and reddish-pink legs. In winter, adult's black neck acquires a white half-collar. Juvenile is similar to adult but has duller legs and a dark-tipped bill. In flight, all birds show a striking white wingbar on otherwise black upperwings (underwings are white with a black trailing edge), and a black-tipped white tail. Utters a loud piping call in alarm. Typically associated with rocky shores but also found on mudflats, especially in winter. Very local breeding species in the region and typically known as a rather scarce winter visitor, present Oct–Mar.

4. DEMOISELLE CRANE *Anthropoides virgo* 85–100cm

Similar to Common Crane but marginally smaller. Note the trailing tertials (bushy in Common Crane) and the black on the head and neck that extends to the upper breast. A close view reveals the white head plumes, white crown and striking red eyes. In flight, the black on the neck extends to the upper breast (with Common Crane, extent of black is limited to neck only). Juvenile is distinguished from juvenile Common Crane by trailing (not bushy) tertials and by association with adults. Typically observed in flight. Resting migrants favour open fields. Scarce passage migrant, mainly in E; may still nest in Morocco.

5. COMMON CRANE *Grus grus* 95–120cm

Large and stately long-legged, long-necked bird. Adult has mainly blue-grey plumage with black and white on face and neck, and red patch on rear of crown; back is sometimes stained buffish brown. Tertial plumes form a bushy 'tail'. Juvenile is similar to adult with mainly grey plumage except that head and neck lack distinct markings and instead are uniform buffish. In flight, all birds show long, broad wings with distinct primary 'fingers'; head and neck are held extended and, in adult birds, appear black from below. Nervous and wary. Forms flocks and flies in V-formation. Feeds in open fields. Utters loud trumpeting calls. Mainly a local winter visitor and passage migrant; Turkish population is resident.

summer

3. winter

1. BLACK-WINGED PRATINCOLE *Glareola nordmanni* 25–28cm

Confusingly similar to Collared Pratincole but distinguishable with care. Note the darker brown upperparts, shorter tail (does not extend beyond the wings in standing birds), the limited amount of red on bill and the darker lores. In flight, the absence of a white trailing edge to the inner wing is a useful identification feature; underwings are all dark and upperwings show less contrast between coverts and flight feathers than in Collared Pratincole. Utters tern-like calls in flight. Favours similar habitats to Collared Pratincole and often associates with that species on migration. Rather scarce passage migrant through E of region.

2. COLLARED PRATINCOLE *Glareola pratincola* 24–27cm

Unusual wader that recalls a plover on the ground but which is tern- or swallow-like in flight. Adult has sandy brown head, neck and upperparts with a white belly; yellow throat is defined by black border. Bill is red at base and black at tip. Juvenile is similar to adult but has scaly appearance on upperparts due to pale feather margins. In flight and from above, adult shows a white rump and forked tail; wings are dark-tipped with a white trailing edge to the inner wing. From below, look for the maroon wing coverts (note, the colour can be difficult to detect) and the white trailing edge to the inner wing. Utters tern-like calls in flight. Typically catches insects on the wing but also feeds on the ground. Favours wetlands for nesting; on migration also found on areas of short grassland. Widespread passage migrant (often in flocks) and locally common breeding species; present Apr–Sept.

3. CREAM-COLOURED COURSER *Cursorius cursor* 24–25cm

Distinctive and unusual plover-like wader of arid terrain. Adult has mainly creamy-buff head, neck, breast and upperparts and white belly; note, however, the dark stripe behind the eye that is separated from the bluish-grey hind crown by a white supercilium. Bill is slender and slightly downcurved and legs are pale flesh in colour. Juvenile recalls adult but the creamy-buff elements of the plumage are scaly, spotted or streaked, and the hind crown is brown not bluish. In flight, all birds show uniformly black underwings and a black outer wing on the upperside. Often runs from danger instead of flying. Favours flat semi-deserts. Resident and summer visitor, mainly N Africa and the Middle East.

4. SENEGAL THICK-KNEE *Burhinus senegalensis* 35–40cm

Similar to Stone-curlew, with streaked brown plumage, yellow legs and staring yellow eyes. Note, however, the relatively larger bill and limited extent of yellow at the base. Standing bird lacks Stone-curlew's black-bordered white wingbar and so, in flight, upper surface of inner wing shows broad, pale patch; black flight feathers show prominent white patches. Favours muddy river margins. Resident, restricted to Egypt's Nile Delta.

5. STONE-CURLEW *Burhinus oedicnemus* 40–45cm

A distinctly secretive and well-camouflaged bird. Essentially nocturnal but occasionally seen feeding at dawn and dusk or flushed from cover in daytime. Its presence is often detected at night by its strange, Eurasian Curlew-like wailing calls. Adult has sandy-brown plumage, yellow legs, black-tipped yellow bill and large yellow eyes. Standing bird shows a black-bordered white wingbar but the dark wings and white wingbars are most apparent in flight. Juvenile is similar to adult but wing markings are less distinct. Favours open terrain with scattered grasses and clumps of vegetation. Common resident in Iberia, NW Africa and parts of the Middle East. Local summer visitor to other parts of the region.

1. NORTHERN LAPWING *Vanellus vanellus* 28–30cm

Well-marked and readily identifiable plover. Adult can appear black and white at a distance. However, at closer range and in good light, shows green sheen to feathers of back; in winter, these have buffish fringes. Spiky crest of feathers is visible at all times, longer in male than female; orange undertail feathers are seen in all birds. Juvenile is similar to winter adult but has shorter crest and even more scaly upperparts. In flight, all birds show rounded, black and white wings and flapping flight. Calls include a strangled *pwee-eh*, and *peeo-wit* at nest site. Favours short grassland and marshes. Forms flocks outside breeding season. Local resident breeder but mainly a winter visitor to the region, present Oct–Mar.

2. RED-WATTLED LAPWING *Vanellus indicus* 33–35cm

Boldly marked and distinctive plover. Adult has black on crown, nape, throat and upper breast; head, neck and underparts are otherwise white except for red eye-ring and lores, and back is brown. Bill is red and dark-tipped and legs are long and yellow. In flight, upperwings show bold brown, white and black pattern and tail is white with sub-terminal black band. Underwings are black and white. Utters distinctive *did-ee-do-it* call. Favours well-vegetated wetlands. Erratic visitor to E of region, mainly in winter, from main range in S Asia.

3. WHITE-TAILED LAPWING *Vanellus leucurus* 27–29cm

Attractive, long-legged plover. Adult has buffish-brown upperparts and neck, a paler greyish-brown face and a grey-brown breast; belly and tail are pure white. Bill is straight and black, and legs are long and yellow. In flight, has distinctive brown, white and black pattern on upperwings and black-tipped white underwings. Juvenile resembles adult but upperparts are spotted and scaly. Favours coastal marshes and wetlands. Formerly a scarce passage migrant in the Middle East but recently has become a regular spring visitor as far W as Greece; in recent years has bred along Romania's Black Sea coast (present May–Sept) and may continue to do so.

4. SOCIABLE LAPWING *Vanellus gregarius* 28–30cm

Attractive plover. Adult has mainly grey-buff or lilac-buff plumage with distinctive facial pattern comprising black eyestripe bordered above and below by white; crown is black and throat is yellow-buff. In summer, adult has black and maroon belly patch; this feature is absent in non-breeding birds. Juvenile resembles non-breeding adult but upperparts are scaly and facial pattern is less distinct. In flight, all birds show distinctive upperwing pattern comprising triangles of lilac-buff, white and black. Calls include a shrill *krrech*. Favours ploughed fields. Scarce passage migrant and winter visitor to Middle East.

5. SPUR-WINGED LAPWING *Vanellus spinosus* 25–27cm

Elegant and long-legged plover. Adult has black and white underparts and a sandy-brown back. Head and neck are white but with a black cap and black band from chin to breast. Tail is white but black-tipped. Bill is dark and straight, and legs are long and black. In flight, shows distinctive grey-brown, white and black upperwings; underwing is white, contrasting with black primaries and belly. Scapular 'spurs' are sometimes visible in flight. Juvenile resembles adult but upperparts and cap appear scaly; 'spurs' on wings are absent. Calls include a shrill *peeik*. Favours muddy saltpans and river deltas. Summer visitor to E Mediterranean, present Apr–Sept. Resident in Egypt's Nile Delta and Middle East.

juv

juv

1

2

3

4

5

1. COMMON RINGED PLOVER *Charadrius hiaticula* 19cm

Dumpy wader. Summer adult has sandy-brown upperparts, white underparts and continuous black breast band and collar. Shows black and white markings on face, and white throat and nape. Legs are orange-yellow and bill is orange and black-tipped. In winter adult, dark elements of plumage are brown; note white supercilium and dark bill flushed orange at base. Juvenile recalls winter adult with subdued dark markings and duller colours. White wingbar seen in all birds. Favours coastal beaches. Winter visitor, present Oct–Mar.

2. LITTLE RINGED PLOVER *Charadrius dubius* 15cm

Smaller and slimmer than Ringed Plover. Summer adult has sandy-brown upperparts and white underparts with black collar and breast band, and black and white markings on face. Dark markings are more intense on male than female. Note also the black bill, yellow legs and yellow eye-ring. In winter, black elements of adult's plumage are replaced by brown. Juvenile is similar to winter adult but has scaly upperparts. In flight, wingbar is absent. Favours shingle riverbeds. Widespread summer visitor, present Apr–Sept; winters in small numbers.

3. KENTISH PLOVER *Charadrius alexandrinus* 15–17cm

Dumpy, pale-looking plover. Adult has sandy-brown back separated from brown cap and nape by white collar; black markings on side of neck form an incomplete collar. In breeding season, male has a chestnut crown and black patch on forehead; at other times, male is similar to year-round appearance of female and crown becomes uniform grey-brown. Juvenile is similar to female but duller. All birds show a wingbar in flight. Essentially coastal, favouring saltpans, brackish lagoons and estuaries. Widespread resident in region.

4. GREATER SAND PLOVER *Charadrius leschenaultii* 20–22cm

Long-legged, large-billed plover. Summer adult male has sandy-brown back and head with black patch through eye, and white forehead and throat. Flushed orange-red from chest, around white throat to supercilium; underparts are otherwise white. Adult female resembles male but lacks black markings and orange colour is reduced to a faint flush. Winter adult recalls summer female but orange-red colour is absent; note the grey-brown shoulder patch, reminiscent of Kentish Plover. Juvenile is similar to winter adult but has scaly upperparts. Favours beaches and pool margins. Passage migrant in E of region.

5. CASPIAN PLOVER *Charadrius asiaticus* 20–21cm

Long-legged plover. Adult summer male has grey-brown upperparts, a dark crown but a white face and throat; note the broad maroon patch on the neck and upper breast, defined and separated from the white underparts by a black line. Summer female and winter adult resemble summer male but lack the reddish breast patch; similar juvenile has scaly upperparts. Favours bare ground. Passage migrant in Middle East; local in NE in summer.

6. EURASIAN DOTTEREL *Charadrius morinellus* 22–24cm

Pot-bellied plover. Summer adult female has reddish-orange breast and belly, separated from blue-grey throat by black-bordered white collar; note the whitish face and white stripe above and behind the eye. Male is similar to, but duller than, female. Juvenile and winter adult have buffish-brown plumage with a hint of the pattern seen in adult birds. Favours short grassland. Passage migrant; also winters in N Africa and Middle East and present there Nov–Mar.

winter

summer

①

winter

summer

②

♂

③

winter

summer

♂

④

♀

⑤

juv

⑥

1. GOLDEN PLOVER *Pluvialis apricaria* 28cm

Robust and plump-bodied plover. Winter-plumage adult and juvenile have spangled golden upperparts and buff underparts marked with bars and streaks. In breeding plumage (seldom seen in region) adult acquires a black belly that grades to grey or marbled black on the face and neck; markings are more distinct in male than female. In flight, all birds show a white wingbar on the upperwings and white underwings. Calls include a whistling *peeoo*. Favours areas of short grass. Widespread but local winter visitor, present Oct–Apr.

2. GREY PLOVER *Pluvialis squatarola* 28cm

Dumpy-bodied plover. Winter adult appears grey, with upperparts spangled black and white and underparts whitish; juvenile is similar but upperparts are suffused with yellow-buff. In breeding plumage (seldom seen in region) adult acquires black underparts, from face and neck to belly. All birds show black 'armpits' (black patch at base of otherwise white underwing) in flight. Call is a distinctive *wee-oo-err*. Favours coastal mudflats. Passage migrant and widespread but local winter visitor, present Oct–Apr.

3. DUNLIN *Calidris alpina* 17–19cm

Familiar long-billed small wader. In winter plumage, adult has uniformly grey-brown upperparts and white underparts. In breeding plumage (seen sometimes in birds in spring and early autumn) has chestnut-brown cap and back, and black belly. Juvenile resembles winter adult but has dark spots on flanks, with grey, black and chestnut on back. Calls include a sharp *priit*. Favours estuaries and mudflats. Widespread winter visitor and passage migrant.

4. CURLEW SANDPIPER *Calidris ferruginea* 19–21cm

Resembles Dunlin but is relatively longer-legged and longer-billed; bill is always markedly downcurved along its length, in the manner of a Eurasian Curlew. Adult in breeding plumage has brick-red coloration to head, neck and underparts; back is well-marked with black, white and red; spring migrants may have incomplete breeding plumage while autumn birds have usually begun to moult. Winter adult has grey-brown upperparts and white underparts. Juvenile is similar to winter adult but has scaly appearance to back and buffish wash to lower neck and chest. All birds show a conspicuous white rump in flight. Favours coastal pools. Widespread passage migrant; overwinters in small numbers.

5. LITTLE STINT *Calidris minuta* 13cm

Recalls a miniature, short-billed Dunlin. Constant frantic activity is a clue to identity. In breeding plumage, adult has white underparts and reddish-brown upperparts with a yellowish 'V' on the mantle; some birds show a striking red flush to head. Winter adult has grey-brown upperparts and white underparts. Juvenile is similar to breeding adult but is paler and has a white 'V' on the back. All birds have dark legs. Calls is a sharp *tsit*. Favours coastal marshes and mudflats. Widespread passage migrant; overwinters in small numbers.

6. TEMMINCK'S STINT *Calidris temminckii* 14cm

Resembles a slim Little Stint and can look (but is not) longer-bodied owing to long wings. Adult has grey-brown upperparts and pale underparts; sharp demarcation between dark breast and pale belly. In summer, has dark centres to many back feathers. Juvenile is similar to adult but has scaly upperparts. All birds have yellow legs. Call is a sharp *tirr*. Favours coastal pools. Widespread passage migrant; overwinters in small numbers (Oct–Apr).

winter

summer

1

summer

winter

2

winter

3. summer

winter

juv

3

summer

4

juv

winter

5. juv

summer

5

6

summer

1. RED KNOT *Calidris canutus* 23–25cm

Robust, medium-sized wader with proportionately short legs and bill. Adult in breeding plumage (sometimes seen in late spring or early autumn) has reddish-orange underparts and grey upperparts spangled with black and red. Winter adult has uniform grey upperparts and pale underparts. Juvenile is similar to winter adult but has scaly upperparts and buff wash to throat and breast. Legs are yellowish in most birds, but adults in full breeding plumage have black legs; bill is mainly dark at all times. In flight, all birds show a striking white wingbar. Call is a sharp *kwuut*. Favours coastal pools and mudflats. Widespread passage migrant across the region; overwinters locally (Oct–Mar).

2. SANDERLING *Calidris alba* 18–20cm

Small and extremely active wader. Distant winter adult looks very white but at close range note the grey upperparts, white underparts, black 'shoulder' patch, and the black legs and bill. Adult in breeding plumage (sometimes seen in late spring or early autumn) has a reddish-brown flush to the head and neck, grey back spangled with black, and white underparts. Call is a shrill *pliit*. Favours sandy beaches and usually seen in small flocks, running at great speed and feeding along edge of breaking waves. Widespread and locally common winter visitor, present Sept–Apr.

3. BROAD-BILLED SANDPIPER *Limicola falcinellus* 16–18cm

Resembles a Dunlin in size and shape but, at most times, has a striking plumage pattern reminiscent of a small Common Snipe. Note in addition the distinctive bill shape: rather straight along most of the length but distinctly downturned at the tip. Adult in summer has streaked brown upperparts and pale underparts with bold streaks on the throat and flanks; head shows a pale supercilium that forks from in front of eye and forms a pale stripe along the side of the crown. Winter adult has mainly grey upperparts and white underparts; still shows the pale, forked supercilium. Juvenile is similar to summer adult but with cleaner, more strongly patterned plumage than observed on adults at that time of year. In flight, shows a white wingbar and a dark-centred pale rump. Call is a buzzing *brr-uut*. Favours muddy pool margins. Scarce passage migrant in E of region.

4. WHIMBREL *Numenius phaeopus* 40–45cm

Superficially similar to Eurasian Curlew but smaller and with proportionately shorter bill and distinctive head pattern comprising two dark lateral stripes on otherwise pale crown. At all times, plumage is essentially grey-brown and heavily streaked. In flight, shows a white wedge-shaped pattern on the rump and lower back, and a barred tail. Often first detected by bubbling call, typically comprising seven rapid whistling notes descending slightly in pitch from start to finish. Favours coastal grassland and shores. Widespread but generally scarce passage migrant. Overwinters in small numbers, in S Iberia and NW Africa.

5. EURASIAN CURLEW *Numenius arquata* 53–58cm

Large and distinctive wader with a long, downcurved bill and long, bluish legs. Bill length is variable and is usually shorter in juvenile than in adult. At all times, plumage is essentially grey-brown and heavily streaked. In flight, shows a white rump and wedge on the lower back; tip of tail has dark, narrow barring. Utters *curlew* call in alarm. Favours coastal wetlands in winter. Widespread winter visitor and passage migrant (Oct–Apr); very local breeder.

summer

2 winter

summer

winter

juv 1

2

4

summer

3

5

1. RUFF *Philomachus pugnax* 24–32cm

Confusingly variable wader both in terms of size and appearance. Male is larger than female but all birds show rather long, yellow legs, a long neck, and a proportionately small head. In spring, adult has grey-brown upperparts and paler, greyish-white underparts, often variably marked with black feathers. In full breeding plumage (seldom seen in region) male acquires extraordinary ruff that can be uniform black, white or reddish orange, or a mixture of colours; note also the bare pinkish-orange skin on the face. Juvenile resembles adult female but back looks scaly and whole body is flushed buffish. In flight, all birds show a white wingbar on the upperwings, pale underwings and a dark-centred white rump. Mostly silent. Favours coastal marshes. Widespread and common passage migrant through the region; overwinters in smaller numbers.

2. BAR-TAILED GODWIT *Limosa lapponica* 38–42cm

Similar to Black-tailed Godwit but shorter legged and with shorter, more upturned bill. In flight, all birds show uniformly dark upperwings, pale underwings and a wedge-shaped white rump and lower back, grading to a narrow-barred tail. Winter adult has curlew-like streaked brown plumage that is darker above than below. Breeding plumage birds are brick-red on the head, neck and underparts, and show a spangled blackish-grey back. Juvenile resembles winter adult but plumage is flushed buffish. Call is a sharp *ker-wee*. Favours coastal mudflats and sand. Winter visitor to W of region, present Sept–Mar.

3. BLACK-TAILED GODWIT *Limosa limosa* 41cm

Large, long-legged wader with a long, slightly upturned bill that is pinkish at the base. In flight, birds in all plumages show a black tail, a white rump and a white wingbar on the upperwings; underwings are white with a dark margin. In winter adult, upperparts are uniformly grey-brown and underparts are pale; juvenile is similar to winter adult but with buffish wash. Summer adult acquires a reddish wash to the head, neck and upper breast, dark barring on the contrastingly pale underparts, and irregular black and reddish markings on the back. Calls include a sharp *aar-ip*. Favours mudflats, estuaries and coastal marshes. Mainly a widespread non-breeding visitor, present Sept–Mar; very local breeder.

4. SPOTTED REDSHANK *Tringa erythropus* 30cm

Resembles Common Redshank but is larger and with proportionately longer red legs and bill. Breeding-plumage birds (occasionally seen in late spring and early autumn) have almost black plumage. Winter adult has grey upperparts, a striking white supercilium and pale underparts. Juvenile is similar to winter adult but with much bolder markings, particularly the barring on the underparts. In flight, all birds have uniformly darkish upperwings, dark-edged pale underwings, and a narrow white wedge on the back. Utters a sharp *tchlewit* call. Favours coastal marshes. Widespread passage migrant. Winters in small numbers (Oct–Mar).

5. COMMON REDSHANK *Tringa totanus* 28cm

Noisy, medium-sized wader that is easily recognised by its red legs and long, red-based bill. Plumage is predominantly grey-brown above and pale below with streaks and barring; it is most heavily marked in the breeding season. In flight, all birds show a broad white trailing edge to the wings and a wedge-shaped white lower back. Presence is often first detected by bird's loud, piping alarm call *teu-hu-hu*. Favours freshwater wetlands and coastal marshes. Local breeder but more widespread as a passage migrant and in winter.

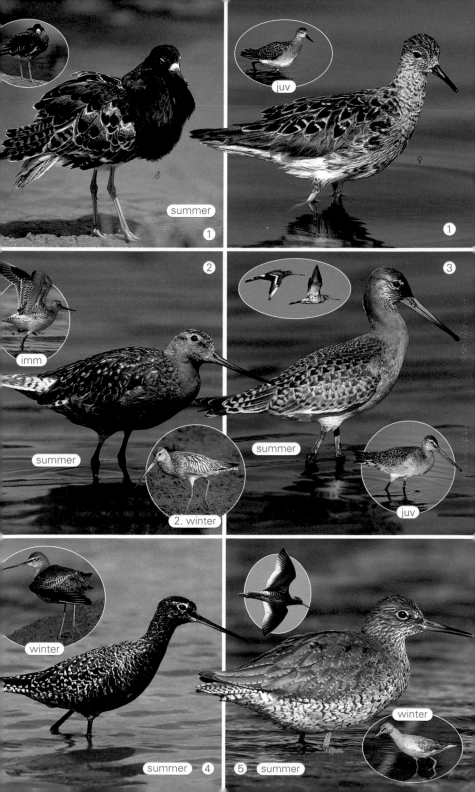

juv

summer

1

1

2

imm

summer

2. winter

3

summer

juv

winter

summer

4

5

summer

winter

1. COMMON GREENSHANK *Tringa nebularia* 30–33cm

Elegant, medium-sized wader with yellow-green legs and a long, slightly upturned bill that is bluish or yellowish at the base. Winter adult is grey above with bright white underparts; shows streaks on the neck and breast. In breeding plumage (seen in late spring), some feathers on back are dark-centred and shows dark spots on neck and upper breast. Juvenile resembles winter adult but upperparts are darker. In flight, all birds show uniform grey upperwings and white wedge from lower back to upper tail. Call is a shrill *tchu-tchu-tchu*. Favours coastal marshes. Widespread passage migrant; winters in small numbers (Oct–Mar).

2. MARSH SANDPIPER *Tringa stagnatilis* 22–25cm

Elegant wader with long legs and a long, needle-like bill. Non-breeding adult has mainly grey upperparts and white underparts; dark shoulder patch contrasts with paler mantle and shows a pale supercilium. In breeding season, upperparts become marked with black and brown and plumage shows black spotting on the neck and upper breast. Juvenile is similar to winter adult but back appears darker and scaly. Call is a whistling *kiuu*. Favours freshwater marshes. Widespread passage migrant, commonest in E; winters in small numbers.

3. WOOD SANDPIPER *Tringa glareola* 20cm

Elegant wader with long, yellow legs and a straight bill. Adult has brownish, spangled upperparts, and well-marked streaking on the head and neck, merging into a pale belly. Juvenile is similar to adult but pale spots on feathers of back are yellowish buff, not whitish. In flight, all birds show a conspicuous white rump; white tail has terminal dark bars that are narrower than on Green Sandpiper. Upperwings are dark and underwings are mostly white. Call is shrill *chiff-chiff-chiff*. Favours freshwater marshes. Common passage migrant; winters in small numbers.

4. TEREK SANDPIPER *Xenus cinereus* 23–25cm

Resembles Common Sandpiper but has a diagnostic long, upturned and yellow-based bill. Legs are yellow, proportionately short and set rather far back on body. Adult has grey upperparts, with a dark 'shoulder' patch and dark scapular stripes; underparts are white. Juvenile is similar to adult but markings on back are less distinct. All birds show a white trailing edge to wing in flight. Call is a shrill *tchu-tu-tu*. Favours coastal mudflats. Passage migrant in E.

5. COMMON SANDPIPER *Actitis hypoleucos* 20cm

Small, plump-bodied wader with an elongated tail end. Adult has warm brown upperparts and white underparts with a clear demarcation between the dark-streaked breast and white belly. Adopts a horizontal stance and bobs up and down when walking. Flies on bowed, fluttering wings and has a conspicuous wingbar. Call is a whistling *tswee-wee-wee*. Favours freshwater habitats. Common passage migrant. Winters in small numbers (Oct–Mar).

6. GREEN SANDPIPER *Tringa ochropus* 23cm

Small, dumpy wader. Adult has dark brown upperparts, spangled with white dots; has heavy streaking on the head and neck with a clear demarcation from the white underparts. Legs are yellowish green and bill is straight and yellow at the base. Feeds quietly along pool margins with a distinctive bobbing gait. When flushed from ditches or ponds, looks black and white with a striking white rump; flight is accompanied by yelping, trisyllabic call *tlewit-wit-wit*. Favours freshwater margins. Widespread passage migrant and winter visitor (Oct–Mar).

winter

1

2

3

1. summer

2

3

3

4

5

6

5

1. EURASIAN WOODCOCK *Scolopax rusticola* 33–35cm

Dumpy wader with rather short legs but a very long bill. Marbled chestnut, black and white plumage affords all birds superb camouflage among dead leaves. Eyes are large and placed unusually high on head. Mainly nocturnal. Favours woodland and scrub. Often only detected when resting bird is flushed by accident during daytime, when it zigzags away on broad wings. Most widespread and common in winter.

2. JACK SNIPE *Lymnocryptes minimus* 19–20cm

Much smaller than other snipe species and has relatively shorter bill. All birds have essentially brown head, neck, breast and upperparts with striking yellowish stripes on back and on head; belly is white. Utters soft *kaatch* when flushed. Superb camouflage makes detection difficult; very reluctant to fly, preferring to 'freeze' until danger passes. Bobs body up and down when walking and feeding. Favours well-vegetated wetlands. Scarce winter visitor (Oct–Mar) and passage migrant.

3. COMMON SNIPE *Gallinago gallinago* 25–27cm

Dumpy, round-bodied wader with rather short legs and an incredibly long, straight bill. All birds have essentially brown plumage on head, neck, breast and upperparts, all patterned with black and white lines and bars; has a white belly and shows dark stripes on head. In flight, upperwings look uniform with a white trailing edge, and splayed tail shows only a limited amount of white. Call is a sneezing *ke-atch*. Feeds by probing downwards with bill in soft mud. Favours freshwater marshes. Mainly a winter visitor (Oct–Apr); very local in summer.

4. GREAT SNIPE *Gallinago media* 27–29cm

Slightly larger than Common Snipe and relatively shorter billed. However, only reliably distinguished using plumage details: note the barred underparts (all except the centre of the belly), three prominent white wingbars (seen both at rest and in flight) and the conspicuous amount of white in the outer tail. Generally silent in the region, although spring migrant males sometimes perform half-hearted clattering and buzzing 'song'. Favours drier habitats than Common Snipe. Scarce and easily overlooked passage migrant in E of region.

5. RUDDY TURNSTONE *Arenaria interpres* 23cm

Dumpy wader with a stubby, triangular bill, used for turning stones. Winter adult and juvenile are variably marked with black, brown and white on upperparts; note the clear demarcation between the dark breast and white underparts. In breeding plumage (seen in late spring), adult has orange-brown feathers on back and striking black and white markings on head. Legs are orange in all birds. In flight, all birds show white wingbar and white on back and rump. Call is a rolling *kutt-tuk-tuk*. Favours coasts. Winter visitor, present Oct–Mar.

6. RED-NECKED PHALAROPE *Phalaropus lobatus* 18cm

Intriguing wader that is typically seen swimming, often spinning rapidly to pick insects off water surface. All birds have a needle-like bill and, on land, lobed toes (that aid swimming) can be seen. Breeding-plumage female has red neck, white throat and mainly dark grey upperparts marked with yellow-buff on back; underparts are white. Male is similar but plumage colours are subdued. Winter adult has grey upperparts, white underparts and black patch through eye. Juvenile is similar to winter adult but upperparts have stronger pattern of black, white and buff markings. Favours coastal pools. Passage migrant in E of region.

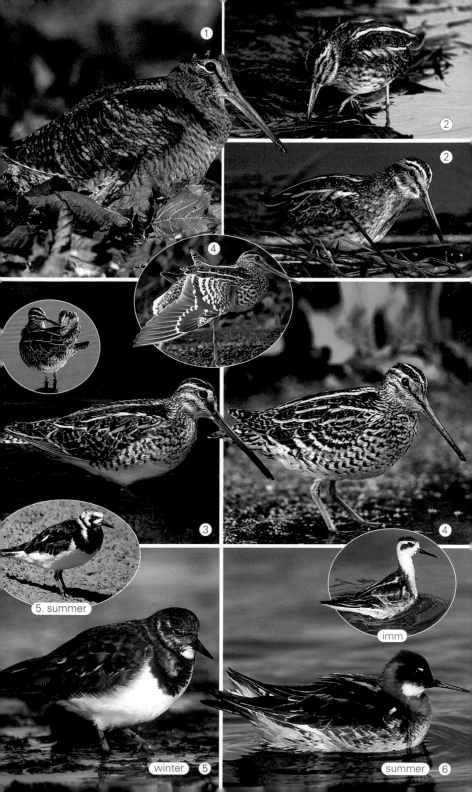

1

2

2

4

5. summer

winter 5

imm

summer 6

1. MEDITERRANEAN GULL *Larus melanocephalus* 37–40cm

An elegant gull, superficially similar to Black-headed but, in all plumages, can be separated from that species with care. In all adult plumages the pure white wings are diagnostic for a bird of this size. In summer, adult has black hood and eyes defined with white 'eyelids'; in winter, dark facial markings are reduced to dark smudges around eyes. First-winter bird is similar to first-winter Black-headed Gull, but dark streaks on head and white 'eyelids' give it a menacing appearance. Second-winter bird is similar to adult but has dark primary tips. Bill is stouter than that of Black-headed Gull; in first-winter birds bill is dark but colour changes with age, becoming blood-red with a dark sub-terminal band in adult. Call is a harsh *cow-cow-cow*. A local breeding species (most numerous in E of region) but widespread on coasts throughout the region in winter. Sometimes gathers in sizeable pre-nesting flocks, particularly around saltpans and other potential feeding grounds.

2. BLACK-HEADED GULL *Larus ridibundus* 35–38cm

A widespread and familiar species. As with other gulls, the plumage varies according to age and time of year. However, at all times it is easily recognised in flight by the white leading edge to the wings. Adult has a grey back and upperwings, white underparts, red legs and a reddish bill. In summer, shows a chocolate-brown hood with white 'eyelids', while in winter the dark facial markings are reduced to dark smudges behind eye. Juvenile has marbled chestnut-brown and grey plumage. First-winter bird is similar to winter adult but grey upperparts show some dark feather centres and colours of bare part are less intense. Calls include a harsh *kree-arr*. Usually found near water (both freshwater and marine) but occasionally feeds on land. A distinctly local breeding species but much more widespread and numerous in winter, especially on coasts.

3. SLENDER-BILLED GULL *Larus genei* 42–44cm

An elegant gull with a bill that is long rather than particularly slender. Most easily confused with winter adult Black-headed Gull, but Slender-billed's elongated forehead and proportionately long neck and legs contribute to the species' attenuated appearance. Summer adult has pale grey mantle, dark primaries and white leading edge to wings. Head, neck and underparts are unmarked white but the latter are often suffused with pink. Legs are pale orange and bill is dark orange-pink (black at height of breeding season). Winter adult lacks pink suffusion of summer bird and often shows grey ear coverts. Juvenile has dirty brown wing coverts and smudges behind eye; legs and bill are pale orange. Calls are similar to, but harsher than, those of Black-headed. Typically favours coastal habitats. A local breeding species of coastal lagoons. Most widespread outside the breeding season.

4. AUDOUIN'S GULL *Larus audouinii* 48–52cm

An elegant gull with relatively long wings and a large bill. Adult has pale grey upperwings and mantle, with a white trailing edge to the wings and black-tipped primaries. Bill is red but with black and yellow tip; legs are black. Juvenile has grey-brown plumage, darkest on the mantle, and dark bill and legs. First-winter bird has a grey mantle, with dark feather centres, mainly dark wings, a dark-tipped bill and a mainly dark tail. Second-winter bird has a reddish bill, grey upperwing and mantle, and a black-tipped white tail. World population is almost entirely restricted to the Mediterranean. Sometimes visits harbours but also occurs far out to sea. Nests on rocky islands; most widespread in winter.

summer 1

2nd-w 1

ad w 1

summer 2

1st-w 2

ad w 2

summer

1st year

winter 3

2nd sum

4

1. YELLOW-LEGGED GULL *Larus cachinnans* 55–65cm

The most common large gull in the region. Adult has a dark grey mantle with white-spotted black wingtips. Bill is yellow with an orange spot near the tip; legs are yellow. During summer months, head, neck and underparts are pure white; in winter, head acquires faint dark streaking on crown and ear coverts. Juvenile is mottled grey-brown with dark bill and legs; acquires adult plumage through successive moults over subsequent two years. Several races occur across region, showing variations in intensity of grey on mantle and extent of white on wingtips. Essentially coastal but will also feed inland on fields and rubbish tips. Local resident breeding species; nests on islands. Widespread in winter. **1a. Armenian Gull** *L. (c.) armenicus* is smaller than typical Yellow-legged; adult has a dark eye and a bill with a sub-terminal dark band and a pale tip. Winter visitor to E.

2. GREAT BLACK-BACKED GULL *Larus marinus* 65–75cm

The largest gull in the region. Adult has a black back and upperwings, with white spots near wingtips, and a broad, white trailing edge to the wings. Head, neck and underparts are pure white at all times. Legs are pinkish and bill is large and yellow with an orange spot near tip of lower mandible. Juvenile has a dark bill and grey-brown, mottled upperparts; head, neck and underparts are white but variably dark-streaked. Adult plumage is acquired in successive moults over the subsequent three to four years. Mainly coastal. Scarce winter visitor (Oct–Mar) to W Mediterranean and Atlantic coasts.

3. LESSER BLACK-BACKED GULL *Larus fuscus* 53–56cm

Similar in size and proportions to Yellow-legged Gull, but mantle and upperwings of adult are darker slate-grey (almost black in some birds); has bright yellow legs and yellow bill with orange spot near tip. By late winter, birds have a pure white head, neck and underparts; in autumn and winter, head and neck are dark-streaked. Juvenile has grey-brown, mottled plumage, darker than that of juvenile Yellow-legged Gull; acquires adult plumage over subsequent two years. Mainly coastal and often found far from land. Widespread and rather common winter visitor, present Oct–Mar.

4. PALLAS'S GULL (GREAT BLACK-HEADED GULL)
Larus ichthyaetus 60–65cm

Large and distinctive gull. Adult has pale grey mantle and upperwings with black, sub-terminal spots on the wingtips; underwing is white with black sub-terminal spots on wingtips. Legs are yellow and bill is stout, grading from yellow at the base to orange but with a dark sub-terminal band and a pale tip. In breeding plumage (seen in late spring), adult has a dark hood with white 'eyelids'; in winter, extent of black is reduced to dark streaks and smudges, through and behind the eye. First-winter bird resembles winter adult but has darker feathering on the upperparts, a black-tipped pink bill and dull pink legs. Acquires adult plumage over subsequent two to three years. Mainly coastal. Winter visitor and passage migrant to the Middle East, present Oct–Apr.

5. LITTLE GULL *Larus minutus* 26–28cm

The smallest gull in the region. Has buoyant, tern-like flight on relatively long wings with rounded tips; those of adult are pale grey above and black below, both sides showing a trailing white margin. In winter, adult has dark smudges on face but, in summer, acquires a dark hood; legs are reddish at all times and bill is dark. Juvenile has striking black bars along the wings and a black-tipped tail. Essentially coastal. Widespread winter visitor, present mainly Oct–Apr.

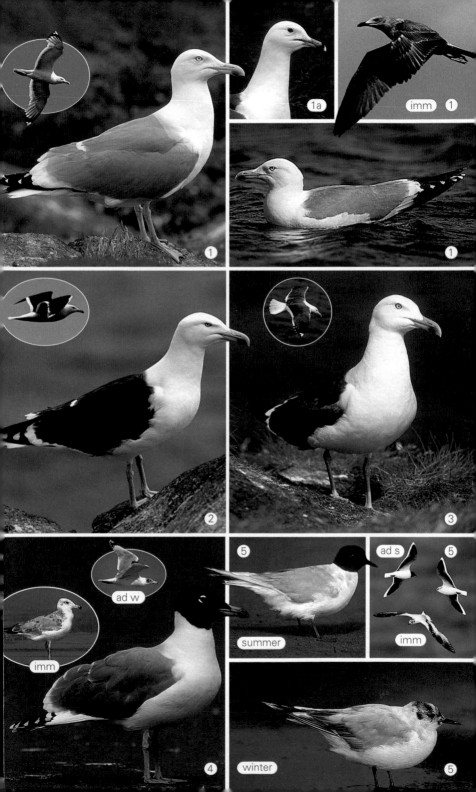

1a

imm 1

1

1

2

3

imm

ad w

4

5

summer

ad s 5

imm

winter 5

1. COMMON GULL *Larus canus* 41cm

Recalls a small version of Yellow-legged Gull. Adult has slate-grey back and upperwings, with white-spotted black wingtips; body plumage otherwise white. In winter plumage (that seen most frequently in region) has dark streaks on back of head and nape. Legs yellowish green and bill yellowish; in winter, bill has dark spot near tip. First-winter bird has bands of grey, brown and black on upperwings. Essentially coastal. Scarce winter visitor (Oct–Mar), particularly to E half of region; easiest to find when winter is hard in N Europe.

2. BLACK-LEGGED KITTIWAKE *Rissa tridactyla* 41cm

A true 'sea' gull, often seen far from land. Flight is buoyant and graceful. Adult has grey back and upperwings, becoming paler towards wingtips, which are tipped black. Body plumage otherwise essentially white, although in winter (when seen in the region) has dark smudges on ear coverts and nape. Bill is yellow and legs are black. First-winter bird has black zigzag on upperwings, a black-tipped tail and dark ear coverts and nape; bill and legs are dark. Invariably found at sea. Winter visitor (Oct–Mar), mainly to W of region.

3. LONG-TAILED SKUA *Stercorarius longicaudus* 40–60cm

Buoyant and graceful, small-bodied skua. Adult has diagnostic long central tail streamers. Shows pale underparts, black cap, grey-brown mantle and dark flight feathers. Juvenile has barred grey-brown plumage, palest on central breast, and lacks tail streamers. Invariably seen at sea and in flight. Chases other seabirds, such as terns, forcing them to regurgitate food. Scarce passage migrant throughout the region. Easily overlooked because of preference for offshore waters.

4. PARASITIC (ARCTIC) SKUA *Stercorarius parasiticus* 46–60cm

Graceful, almost falcon-like seabird. Adult has narrow pointed wings with a white patch near the tips and pointed tail streamers projecting beyond the wedge-shaped tail; occurs in two colour forms, one essentially all dark, the other similar but with pale underparts and dark cap. Juvenile has barred brown plumage and lacks projecting tail streamers. Invariably seen at sea and in flight. Chases other seabirds, such as terns and small gulls, forcing them to regurgitate food. Passage migrant throughout the region.

5. POMARINE SKUA *Stercorarius pomarinus* 65–75cm

A bulky, plump-bodied skua. Adult has diagnostic twisted, projecting tail streamers. Most birds have pale underparts and nape, with a dark cap and complete or incomplete sooty breast band; plumage is otherwise dark brown but shows a white patch near the wingtips. Dark-phase adults are much more unusual, with plumage all dark except for white wing patches. Juvenile has barred brown plumage and lacks tail streamers. Invariably seen at sea and in flight. Chases other seabirds, forcing them to regurgitate food. Passage migrant throughout region.

6. GREAT SKUA *Stercorarius skua* 54–60cm

Bulky and powerful seabird. Adult has mottled brown plumage that resembles that of an immature gull; bill and legs are dark. In flight, looks proportionately broad-winged and short-tailed; shows a white patch near the wingtip. Juvenile is similar to adult but overall plumage is darker. Usually seen in flight and at sea. Chases seabirds the size of gannets, forcing them to regurgitate food. Winter visitor and passage migrant, mainly to W of region.

1. ad w   1. 1st-w   2. ad w

3

4

1

5

2

6

2. juv

1. SANDWICH TERN *Sterna sandvicensis* 38–41cm

Similar to Gull-billed Tern but distinguishable by its long and narrow, yellow-tipped dark bill, its shaggy crest and its short, dark legs. Easily recognised in flight by its powerful, buoyant wing action on long, narrow wings; also identified at a distance by its frequently uttered harsh *churrick* call. Back and upperwing of summer adult are pale grey but plumage is otherwise white except for the dark crest; in winter plumage (sometimes seen in autumn birds), loses the dark cap but retains black on the nape. Juvenile has a scaly-looking back, a dark brown cap and an all-dark bill. Common passage migrant and local breeding species, present mainly Apr–Sept; winters in small numbers.

2. GULL-BILLED TERN *Sterna nilotica* 35–38cm

Bulky tern with a uniformly dark, robust and gull-like bill and rather long, dark legs. In breeding season, adult has a black crown with white face and underparts; upperparts are grey except for the dark primary tips. Winter adult loses the black cap but retains a dark mask through the eye. Juvenile is similar to winter adult but shows brown feathering on the grey upperparts. In flight, adults look pale winged except for the dark trailing edge to the primary feathers. Call is a harsh *chrr-vick*. Favours open wetlands, mainly in coastal areas on migration. A passage migrant and summer visitor to the region with scattered breeding colonies, present May–Sept.

3. COMMON TERN *Sterna hirundo* 35cm

Elegant and graceful medium-sized tern. Adult has a pale grey back and upperparts, and otherwise white plumage. Legs are red and bill is orange-red with a black tip. Black cap is present in summer adult but incomplete in winter plumage (sometimes seen in autumn birds). Juvenile is similar to winter adult but has scaly appearance to back and dark leading edge to inner wing. Calls include a clipped *kip* and a harsh *kee-errr*. Favours both coastal seas and freshwater lakes. Mainly a widespread passage migrant and local breeder, present May–Sept; small numbers may overwinter in far W of region.

4. LESSER CRESTED TERN *Sterna bengalensis* 35–40cm

Slim and elegant tern. Summer adult has grey back and upperparts; plumage otherwise white except for black cap and shaggy crest. Bill is long, slender and orange-yellow, and legs are short and black. Winter adult is similar, but black on head is reduced to area behind the eye. Juvenile resembles winter adult, but feathers on back and upperwings are marked by brown and black. Call is similar to that of Sandwich Tern. Favours coastal seas. Breeds (Apr–July) very locally on N African coast and occasionally attempts to nest elsewhere in the region; otherwise known as a passage migrant.

5. CASPIAN TERN *Sterna caspia* 47–54cm

Huge and impressive tern, the size of a Yellow-legged Gull and with a massive, blood-red bill. Legs are black. Adult has grey mantle and upperwings and white underparts. During breeding season, has black crown and nape but, in winter, dark cap is incomplete and also bill colour fades. Juvenile resembles winter adult but back looks scaly and legs are dull pink. Flight is powerful and buoyant in all birds. Call is a loud and rasping *kree-ahr*. Favours sheltered coastal waters with sand banks and islets for nesting and roosting. Widespread but scarce passage migrant. Known mainly as a local winter visitor and scarce passage migrant but breeds (May–July) very locally and in small numbers in E.

winter

summer 1

summer 2

summer 3

summer 4

summer 5

1. LITTLE TERN *Sterna albifrons* 24cm

The smallest *Sterna* tern in the region, similar in size to species of *Chlidonias* terns (the so-called marsh terns). Easily recognised by combination of pale plumage and small size alone. In breeding plumage, adult has pale grey back and upperwings with black tips to outer primary feathers; tail and underparts are white and crown is black except for white fore-crown. Black-tipped yellow bill and yellow-orange legs are obvious and diagnostic. In non-breeding adult plumage (acquired from August onwards), shows more white on forecrown and between eye and bill; bill is black and leg colour darkens. Juvenile is similar to non-breeding adult but has scaly feathering on back. Flight is buoyant and birds frequently hover before plunge-diving for small fish and crustaceans. Feeds on coasts, coastal lagoons and rivers. Breeds on undisturbed shingle and sandy beaches, but also beside rivers. A widespread summer visitor, present Apr–Sept.

2. BLACK TERN *Chlidonias niger* 24cm

Adult in breeding plumage (Apr–June) has black body, dark grey wings and rump, and white stern and tail. From midsummer onwards, body plumage of adult becomes mottled with pale feathers and eventually (by Sept) becomes pale grey on upperparts, including rump, and white on head and underparts except for black on nape and crown; grey on back extends to dark patch on side of breast. Bill is black and legs are dull red at all times. Juvenile is similar to non-breeding adult but back feathers have pale margins; tail and rump are grey. Often seen in flocks, especially in spring, feeding over coastal marshes and catching insects on the wing. Common passage migrant through the region and a local breeding species; present Apr–Sept.

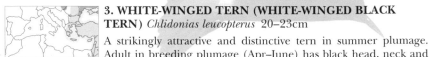

3. WHITE-WINGED TERN (WHITE-WINGED BLACK TERN) *Chlidonias leucopterus* 20–23cm

A strikingly attractive and distinctive tern in summer plumage. Adult in breeding plumage (Apr–June) has black head, neck and body, including a dark mantle. Upperwings show a striking white forewing patch and grey flight feathers, while underwings have black coverts and pale grey flight feathers. Rump, stern and tail are pure white. From midsummer onwards, plumage becomes mottled with pale feathers, and by Sept upperparts are grey (palest on forewing panel and tail) with contrasting white rump; underparts are white and head shows grey on nape and black on ear coverts. Bill is dark and legs are red at all times. Juvenile is similar to non-breeding adult but has a dark back. Often seen in flocks, mixed with other marsh terns, hawking insects over wetlands. Common passage migrant in E Mediterranean but rather scarce elsewhere; seen mainly Apr–May and Aug–Sept, but breeds very locally in NE of region.

4. WHISKERED TERN *Chlidonias hybrida* 23–28cm

An attractive marsh tern that resembles a miniature and compact Common Tern but is distinguished by its smaller size and less forked tail. Adult in breeding plumage (Apr–July) has black crown and nape and striking white cheeks; plumage is otherwise smoky grey, darkest on throat and belly, and palest on silvery-white flight feathers. In non-breeding plumage (acquired by Sept), upperparts are pale grey, except for dark streaking from rear of crown to nape, while underparts are white. Legs and bill are red, the colour most intense in breeding plumage. Juvenile is similar to non-breeding adult but has a dark brown mantle contrasting with a whitish rump and pale grey wings. Typically a widespread passage migrant through the region and local breeder; present Apr–Sept.

1

2. non-br

2

3

4. non-br

4

3. non-br

1. PIN-TAILED SANDGROUSE *Pterocles alchata* 30–40cm

Plump, partridge-like bird with long central tail feathers. Colourful male has olive-green crown, neck and back, with orange on face and orange breast band defined above and below by black line. Note the black chin and dark eyestripe. Belly is white, while back is adorned with pale spots and wing coverts are barred. In flight, shows striking white underparts, including wing coverts. Female resembles male but colours on face, neck and breast are more uniformly dull orange; has two narrow black bands on neck. Favours arid, sandy country and visits drinking pools in mornings. Flight is direct and powerful. Call is a rasping *krrr-rrr*, delivered in flight. Forms flocks outside the breeding season. Locally common resident in Iberia, N Africa and Middle East.

2. BLACK-BELLIED SANDGROUSE *Pterocles orientalis* 30–35cm

Dumpy, partridge-like bird with distinctive markings. Male has blue-grey head and neck, throat flushed with orange, grading to black in centre. Breast is pale buff, defined above by a black line, and belly is black. Back and upperwings are marbled brown. In flight, pale 'armpits' and underwing coverts contrast with the black flight feathers. Female is similar to male, but head, neck and breast are dark-streaked buff, and back has intricate dark markings. Calls include a trilling *hrrrr*, delivered in flight. Favours arid terrain, including stony plains, sometimes at comparatively high altitudes; visits waterholes in mornings. Forms flocks outside breeding season. Resident in Iberia, N Africa, Middle East and from Turkey eastwards.

3. SPOTTED SANDGROUSE *Pterocles senegallus* 30–35cm

Wary, partridge-like bird with long central tail projections. Narrow black line on belly is easily overlooked and usually most obvious in flight. Male has pale blue-grey head, neck and breast, except for orange on face and throat. Underparts (including underwing coverts) are pale sandy-buff, while back and upperwings are buff with grey bands and pale spots. Female resembles male but upperparts and neck are extensively marked with fine dark spots. Favours semi-deserts and arid plains. Call is shrill *kwitt-oo*, delivered in flight. Visits waterholes in mornings. Forms small flocks outside breeding season. Widespread resident in N Africa and Middle East.

4. LICHTENSTEIN'S SANDGROUSE *Pterocles lichtensteinii* 23–26cm

Partridge-like bird with little tail projection. Male has dense pattern of dark vermiculations on most of body but shows an orange-buff breast band. Note the striking black and white bars on the forecrown. Female is overall buffish brown with dense vermiculations. Underwing is uniform grey. Visits pools after sunset. Calls include a sharp *kwitt-al*. Local resident in Middle East and NW Africa.

5. CROWNED SANDGROUSE *Pterocles coronatus* 26–30cm

Male has essentially sandy-buff plumage but striking head pattern with orange-buff on face, whitish around eye, a reddish crown bordered below by blue-grey, and a vertical black line at the base of the bill. Back and upperwing coverts are overall dark brown but are adorned with pale spots. In flight, pale underparts contrast with dark flight feathers. Female resembles male but is more extensively marked with dark spots, particularly on the breast; also lacks the male's black vertical facial line. Favours semi-deserts and arid plains. Visits pools in mornings. Calls include a rattling *chakka-chakka-ka*. Flight is fast and direct, recalling Golden Plover. Forms flocks outside breeding season. Local resident in N Africa and Middle East.

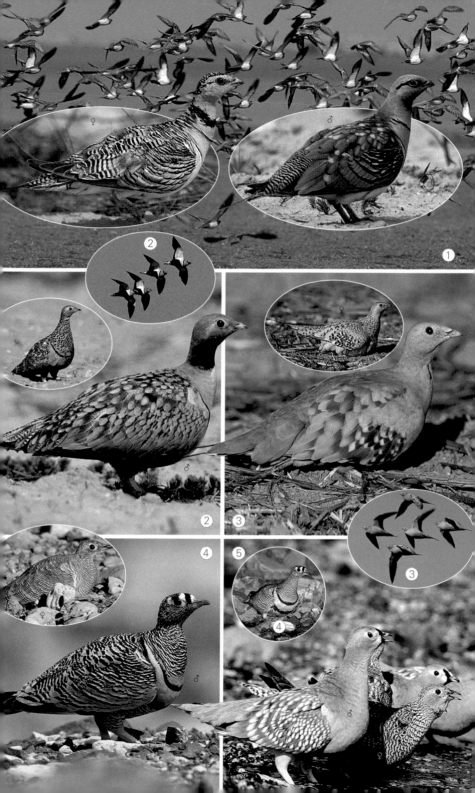

1. ATLANTIC PUFFIN *Fratercula arctica* 30–33cm

Small, dumpy seabird, invariably seen either swimming far away from land or flying over sea on whirring wings. In winter plumage (that seen in the region) has mainly black upperparts and white underparts. Head has black cap and dark grey face, while bill is parrot-like with zones of dull red, yellow and blue. Winter visitor (Nov–Mar) to W of region.

2. RAZORBILL *Alca torda* 40–42cm

Robust and dumpy seabird seen either swimming on sea or flying low over water. In winter plumage (that seen in the region) has black upperparts and white underparts. Bill is deep, hook-tipped and black with a white vertical line; head shows a dark cap and white face. Winter visitor to W of region, present Nov–Feb.

3. COMMON WOODPIGEON *Columba palumbus* 41cm

Plump-bodied pigeon. Adult has blue-grey head, neck and upperparts, with pinkish maroon on the breast, grading to white on the belly. Note the white patch on side of neck and, in flight, prominent, transverse white wingbars. Tail is rather long and dark-tipped. Eye has yellow iris and bill is yellow, pink and white. Juvenile resembles adult but lacks white neck markings and has a dull bill. When disturbed, all birds fly off with loud clattering wings. Song is a series of *oo-OO-oo, oo-oo* phrases. Feeds on farmland but nests in wooded areas. Forms flocks outside breeding season. Widespread resident from Iberia to S Turkey, and NW Africa; numbers boosted and range extended E in winter by migrants.

4. STOCK DOVE *Columba oenas* 33cm

Resembles Woodpigeon. Adult has uniform blue-grey upperparts with two narrow black wingbars on upper surface of inner wing and a patch of iridescent blue feathers on side of neck. Bill is yellow, red and white, while eye is dark. Juvenile recalls adult, but note the dark bill and uniform blue-grey neck. Favours open agricultural land with scattered woodland. Feeds in flocks and sometimes associates with Woodpigeons where the feeding is good. Nests in tree holes and, during breeding season, utters diagnostic and repetitive *ooo-look* call. Locally common resident; more numerous and widespread in winter due to migrant visitors.

5. EURASIAN COLLARED DOVE *Streptopelia decaocto* 32cm

Has sandy-brown plumage, flushed pinkish on head and underparts. Note the dark half-collar on the nape. Black wingtips and white outer tail feathers are best seen in flight. In display, glides on bowed wings. Repetitive song comprises *oo-OO-oo* phrases. Favours human habitation, and often found in villages. Typically seen in pairs and almost never in flocks of any size. Range has expanded in recent decades, and continues to do so.

6. ROCK DOVE (including FERAL PIGEON) *Columba livia* 33cm

Rock Dove is the wild ancestor of the familiar Feral Pigeon. A true Rock Dove is recognised by its blue-grey plumage, two broad black wingbars, white rump and black-tipped grey tail. Some Feral Pigeons show ancestral-type plumage but most show a spectrum of additional or alternative colours and features. Utters a subdued *droo-ooo*. Rock Dove is a locally common resident of rocky coasts and rugged, interior terrain while Feral Pigeon is common in urban areas. Feral Pigeons are often tame while Rock Doves are shy and unapproachable.

winter

winter

winter ①

winter ②

③

④

⑤

⑥

1. LAUGHING DOVE *Streptopelia senegalensis* 25–26cm

Small and rather slim dove with a subtle range of colours in its plumage. Head, neck and breast are pinkish buff to lilac-buff and underparts are otherwise whitish. Back and upperwing coverts are orange-brown; this colour is separated from the dark wingtips by a wedge of blue-grey (this forms the lower edge of the wing in resting birds). Tail pattern is distinctive in flying birds: greyish centre is flanked by white outer corners and a dark spot on either side at the base. Call is a soft *oo-oo-OO-OO-oo*. Favours areas of lush, tropical vegetation in warm climates; typically found at oases in arid areas. Resident in N Africa and the Middle East and, more locally, N into Turkey.

2. EUROPEAN TURTLE DOVE *Streptopelia turtur* 26–27cm

Attractively marked little pigeon with the proportions of a Collared Dove but smaller. Body plumage is mostly blue-grey and pinkish; note the contrasting chestnut-brown feathers on the mantle that have dark feather centres resulting in a scaly appearance to the back. Tail is long and mainly black; appears wedge-shaped in flight due to striking, broad white corners. At close range, black and white barring on the neck can be seen. Presence is often indicated by soft, purring song. Favours orchards and agricultural land with scrub. Common passage migrant and widespread breeding species, present May–Aug.

3. GREAT SPOTTED CUCKOO *Clamator glandarius* 38–40cm

Resembles Common Cuckoo but is larger and has a distinct crest, broader wings and a longer tail. Adult has a dark grey crest (often rather sparsely feathered); upperparts are otherwise dark grey-brown with distinct white tips to the feathers on the back, wings and tail. Underparts are white with a buffish flush to the throat. Juvenile lacks adult's prominent crest, and its plumage has a browner hue; feather tips are pale yellow and has chestnut primaries that are particularly striking in flight. Calls include a range of harsh, chattering sounds, including a coarse *kree-kree-kree*. Favours open woodland. Nest parasite of the crow family. Resident in S Spain but elsewhere typically a passage migrant and summer visitor, mainly to Iberia, S France, N Italy and Turkey; present Apr–Aug.

4. SENEGAL COUCAL *Centropus senegalensis* 35–40cm

Large, cuckoo-like bird with a proportionately long, dark tail and a large, black bill. Adult has a black cap, grading to dark grey-brown on the nape and back. Wings are reddish brown and underparts are creamy-buff; eye has a red iris. Juvenile is similar to adult but back is barred, cap is grey not black, and iris is pale. Calls include a bubbling *OO-OO-oo-oo-oo*. Favours areas of scrub surrounded by lush vegetation. Usually feeds in dense cover and generally rather secretive. Resident in Egypt.

5. COMMON CUCKOO *Cuculus canorus* 33cm

Best known for the male's familiar call. Adult male and most females have a grey head and upperparts, and barred white underparts; legs, iris and base of bill are yellow. Juvenile and some females have brown, barred plumage; juvenile has a pale nape. Male's familiar *cuck-oo* call is heard more often than the bird itself is seen during first six weeks after arrival in late Apr; female utters a bubbling call. Secretive but sometimes perches on fenceposts. In low-level flight, resembles a Sparrowhawk and often has the same alarming effect upon small birds. Nest parasite of songbirds. Widespread passage migrant and breeding species, present Apr–Aug.

①

②

③

④

⑤

juv

1. BARN OWL *Tyto alba* 34cm

A distinctive and pale owl with a unique heart-shaped white facial disc and dark eyes. Flying bird typically looks strikingly pale when seen after dark, caught, for example, in car headlights. Birds from most parts of the region (including the Middle East) typically have pale orange-buff upperparts, marbled with grey and speckled with tiny black and white dots; underparts and underwing coverts are white. Birds from NE of region have darker upperparts with orange-buff underparts and underwing coverts. Typical call is a piercing and blood-curdling shriek. Favours open agricultural land and grassland with woodland and scrub. Mainly nocturnal but sometimes seen on the wing at dusk. Flight is leisurely and slow on wings that are long and rounded at the tip; typically quarters the ground, doubling back when prey is detected and plunging down with dangling legs. Specialises in catching small mammals but will also take insects and small birds. Often nests in buildings and can become rather indifferent to human observers when breeding in villages or rural settlements. Widespread and relatively common resident in W and central parts of the region, but scarce in E.

2. TAWNY OWL *Strix aluco* 38cm

A widespread and strictly nocturnal owl. Specialises in catching small mammals but will also take amphibians and insects. Roosting birds typically look plump and rather rounded. Plumage is overall chestnut-brown but variation in ground colour occurs (some birds are grey-brown while others are rufous). Streaked underparts look greyish and paler than the upperparts, which are well marked with dark streaks. Note the rounded head and facial disc and the black eyes. Underwings are barred greyish white and so can look surprisingly pale when seen in flight from below and caught, for example, in car headlights. Pale, rather fluffy young birds often leave the nest before they are fully independent and fledged. Roosts during the day among foliage of trees and is generally difficult to locate. Utters a sharp *kew-ick* call but best known for the male's familiar hooting calls. Typically favours broadleaved woodland, but also found in surprisingly open and comparatively tree-less terrain so long as scattered bushes are present. Widespread and relatively common resident throughout most of S Europe; more local in S Turkey, the Middle East and NW Africa.

3. EURASIAN EAGLE OWL *Bubo bubo* 60–75cm

A huge and dumpy owl, and the largest of its kind in the region. Capable of catching prey the size of a brown hare, and occasionally even larger. Across most of the region, birds have mainly grey-brown upperparts with dark feather markings (streaking and marbling); underparts are buffish brown with dark streaks and intricate soft vermiculations. Head has prominent ear tufts and large, orange eyes. Chin and throat have pale feathering and legs and feet are feathered too. Birds from the Middle East are generally smaller and overall much paler buff, the darker feather markings showing less contrast. All birds look impressive and vaguely buzzard-like in flight, but with proportionately much larger head and neck. Roosts unobtrusively during daytime on cliff ledges or among dense conifer foliage and, despite its size, is easily overlooked. Usually takes to the wing just after sunset. Its deep, booming *boo-hoo* call is sometimes heard at dusk. Favours cliffs and gorges, and is generally intolerant of continued disturbance. Widespread but scarce resident in S Europe, NW Africa, Turkey and the Middle East.

1

2

2. juv

Middle East

3

1. SHORT-EARED OWL *Asio flammeus* 38cm

A medium-sized and well-marked owl that often hunts in the day-time. Upperparts typically appear overall buffish brown but heav-ily marked with dark and pale spots. Underparts are pale buff, suffused orange-buff on the neck and upper breast and marked with dark streaks, these most intense on the neck and upper breast. Note the pale, rounded facial disc, short 'ears' (often not visible) and the staring yellow eyes. Flight is buoyant and slow on long, round-tipped wings. Typically, wings are held stiffly and flight is jerky; frequently glides. Underwings look pale but note the dark tips. Favours areas of short grassland. Sometimes seen perched on the ground or on fenceposts. Widespread and generally scarce winter visitor to S Europe; much more local in NW Africa and the Middle East; present Nov–Mar.

2. LONG-EARED OWL *Asio otus* 36cm

Superficially rather similar to Short-eared Owl but strictly noc-turnal. Plumage is overall darker than that of Short-eared Owl, with intricate dark markings on the upperparts and finer streaking on the underparts. Rounded facial disc is orange-buff and the eyes are orange. Raises its relatively long 'ears' when agitated or alert; at these times whole body becomes elongated too. Seen from below, flight feathers are pale but wing coverts are orange-buff; wingtips are rounded. Compared to Short-eared Owl, note the barred, pale wingtips. Outside the breeding season, often roosts communally, usu-ally favouring dense conifer foliage; roosts can sometimes be detected by the col-lections of pellets that litter the ground. Favours areas of woodland in open country. Widespread resident in S Europe; much more local in NW Africa and the Middle East. Numbers boosted in winter by migrant visitors.

3. LITTLE OWL *Athene noctua* 22cm

The region's most frequently seen owl, owing to its partly diurnal habits but also because generally it is quite common. Readily iden-tified by its rather small size and rounded appearance. Across most of the region, the plumage is mainly grey-brown. Close inspection reveals the upper-parts to be marked with large white spots while the underparts are streaked. Birds from desert regions of the Middle East and N Africa are extremely pale sandy-brown. Characteristic of all birds are the staring yellow eyes that glare back at observer. Perches on fenceposts, boulders and dead branches, often bobbing head and body when agitated or alert. Typically favours agricultural land and often nests and roosts in buildings. Calls include cat-like *kiu*. Widespread throughout the region.

4. EUROPEAN SCOPS OWL *Otus scops* 19–20cm

A small, well-camouflaged and strictly nocturnal owl with conspic-uous ear tufts (although these are not always clearly visible). Occurs in two colour forms: either overall grey-brown or rufous. Close inspection reveals both colour forms to have a delicate pattern of bars and streaks on the plumage and a striking line of black-margined white spots on the scapulars; the pattern and markings create a camouflaged effect resembling tree bark. The eyes are yellow but they are often closed in roosting birds. Territorial male utters a monotonous sonar-blip call throughout night. Favours woodlands and olive groves. Roosting birds sit tight and are difficult both to locate or to flush. Widespread summer visitor to S Europe, Turkey and NW Africa, present Apr–Sept, although partial resident in southernmost Europe. Passage migrant through the Middle East and N Africa.

1. EUROPEAN NIGHTJAR *Caprimulgus europaeus* 27cm

Nocturnal habits make observation hard. Cryptic plumage affords it good camouflage when resting on the ground: brown, grey and black markings resemble wood bark. Sits motionless, even when approached closely. At dusk, takes to the wing and hawks insects. Raptor-like in flight, with relatively long tail and narrow wings; male has white on wings and tail. Male utters churring song for hours on end at night. Found in open conifer woodland, maquis and on heaths. Widespread summer visitor (May–Aug). **1a. Egyptian Nightjar** *C. aegyptius* is

paler, sandy brown and lacks white markings on wings and tail. Very local summer visitor and migrant to N Africa and Middle East.

2. RED-NECKED NIGHTJAR *Caprimulgus ruficollis* 30–32cm

Strictly nocturnal bird. Larger than European Nightjar and with longer wings and tail. Adult mainly rufous-brown and grey, marked with pale spots and pale line on scapulars. Note the buffish-orange throat and collar, and white 'moustache' and chin. In flight, both sexes show white on wings and outer tail feathers. Favours Stone Pine woodlands and dunes. Song a repeated *kutoc-kutoc*. Summer visitor to Iberia and NW Africa, present Apr–Oct.

3. ALPINE SWIFT *Apus melba* 20–22cm

Large and distinctive swift. Invariably seen in flight, when bulky body and crescent-shaped wings are apparent. In good light, adult looks sandy-brown above. Seen from below, the white throat and white belly, separated by a dark collar, are key features; the underparts otherwise appear dark. Nests on cliffs and buildings, and feeds around coastal cliffs and mountains. Twittering calls often attract observers' attention. Summer visitor, present Apr–Sept.

4. PALLID SWIFT *Apus pallida* 16–17cm

Plumage overall sandy-brown with paler forehead and throat. Flight feathers can look translucent against the light. Compared to Common Swift, has proportionately bulkier body, broader wings and shorter, more rounded tail forks. Favours villages and coasts. Screaming call is similar to that of Common Swift. Widespread summer visitor, present Apr–Sept.

5. COMMON SWIFT *Apus apus* 16–17cm

The commonest swift in the region. Has anchor-shaped outline in flight and all-black plumage except for paler throat and forehead. Tail is more deeply forked than that of Pallid Swift; this feature is not always apparent since forks are often held closed. Extremely vocal; parties of swifts are often heard screaming as they chase one another through streets or over rooftops. Favours towns and villages. Summer visitor throughout, present May–Aug.

6. LITTLE SWIFT *Apus affinis* 12–14cm

Small, compact swift, reminiscent of House Martin, with relatively short, square-ended tail. Plumage is mainly dark except for broad white rump and white throat. Nests in roofs and cliff holes. Resident and summer visitor, mainly NW Africa and Middle East; has also nested in S Spain.

7. WHITE-RUMPED SWIFT *Apus caffer* 15cm

Small, dark swift with a narrow white rump, a white throat and a proportionately long, deeply forked tail; pale tips to secondaries visible only at close range. Usually seen in flight; favours rugged uplands. Local summer visitor to NW Africa and very local in S Spain (May–Sept).

1a

1

2

adult & chick

2

6

3

7

5

3

4

1. EUROPEAN BEE-EATER *Merops apiaster* 27–29cm

The quintessential Mediterranean bird, whose bubbling *pruuupp* calls are such a feature of the months of May–Aug. Has amazingly colourful plumage. Adult has a chestnut crown and nape, grading to yellow on the back and rump; uppertail is green with two projecting central feathers and underparts are blue except for black-bordered yellow throat. In gliding flight, wings show chestnut and blue on the upper surface. Close views of perched birds reveal a dark mask through the eye and a white forecrown. Juvenile is similar to adult but colours are duller and tail lacks central projections. Often seen flying in flocks, circling and gliding in pursuit of insects. Usually returns to perch with prey and removes stings from bees and wasps by knocking them against perch. A colonial nester, excavating holes in sand banks. Hunts over agricultural land, marshes and rivers. A widespread and common summer visitor, breeding throughout the region. Migrating flocks often fly high and are easier to hear than to see.

2. LITTLE GREEN BEE-EATER *Merops orientalis* 21–25cm

A small and delicate-looking bee-eater, appreciably smaller than its European or Blue-cheeked cousins. Shows black eyestripe and black border line to lower throat. Otherwise plumage is mainly green, most intense in subspecies *M. o. cyanophrys* from Israel. This subspecies also has a bright blue throat and supercilium, and proportionately shorter tail projections than the subspecies *M. o. cleopatra* from Egypt. Juveniles are similar to adults of respective subspecies. Local resident in N Nile Valley and Israel. Favours dry, open country with scattered bushes.

3. BLUE-CHEEKED BEE-EATER *Merops persicus* 28–32cm

In silhouette, superficially similar to European Bee-eater but a good view reveals a proportionately longer bill and longer tail projections. Adult plumage is mainly green but shows a dark eyestripe and a colourful throat that grades from red to yellow. Cheeks are pale blue with pale (almost white) border above and below the dark eyestripe. In flight, looks mainly green from above, but from below the reddish underwings contrast with the green body. Juvenile is similar to adult but plumage colours are duller (especially on throat) and tail lacks central projections. A summer visitor (mainly May–Sept) to the Middle East and, more locally, to NW Africa; favours dry, open country. Elsewhere, seen as a scarce vagrant or passage migrant, mainly in spring and sometimes in the company of European Bee-eaters.

4. EUROPEAN ROLLER *Coracias garrulus* 30–32cm

A colourful and distinctive bird of crow-like proportions and size. Adult has blue head, neck and underparts, palest on forehead, and shows a dark patch through the eye. The back is chestnut and the rump and tail are bluish purple. In flight, the wings look striking, the dark blue flight feathers contrasting with pale blue coverts and darker shoulder. The legs and feet are dark. Juvenile is similar to adult but with duller plumage. Favours dry, open country with scattered trees. Often perches on wires or dead branches and scans the ground below for lizards and large insects; these are caught and dispatched using the robust, hook-tipped bill. A widespread passage migrant throughout most of the region. Also a locally common breeding species, nesting in tree holes and also in abandoned buildings. Present in the region mainly May–Aug.

1. EURASIAN HOOPOE *Upupa epops* 26–28cm

A distinctive bird with pale pinkish-brown plumage, except for striking black and white barring on wings. Intensity of pinkish colour varies between individuals and some birds are extremely pale. Unobtrusive when feeding on ground, especially since it often adopts a crouching posture and favours furrows and natural depressions in ground. Striking and unmistakable in flight, with broad black and white wings and slow, butterfly-like flight pattern. White rump visible as bird flies away. Long, downcurved bill is used to probe ground for invertebrates. Erectile crest of barred pink feathers is raised in excitement. Favours open agricultural land and areas of short grass. Nests in tree holes and in holes in stone walls. Utters diagnostic *hoo-poo-poo* call. Widespread and common summer visitor to warm, dry land bordering the N shores of the Mediterranean and present Mar–Sept. Occurs year round in S Iberia, NW Africa and parts of the Middle East; elsewhere seen as a passage migrant.

2. WHITE-THROATED KINGFISHER *Halcyon smyrnensis* 26–29cm

Medium-sized kingfisher with beautiful plumage. Adult has chestnut head, neck and belly with a dazzling white throat and upper breast. Back, tail and wings are iridescent blue but with darker wing coverts and black tips to primary feathers. Has large, bright red bill and red legs. Juvenile is similar to adult but with duller colours. Favours wetlands with open water and usually breeds in riverbanks. Usually perches over water but will also be found on adjacent areas of dry land, where it feeds on small lizards and insects such as mole crickets. Utters rattling and whistling calls when excited. Resident in small numbers in E Mediterranean, particularly along S coast of Turkey and Middle East; occasionally wanders in winter.

3. PIED KINGFISHER *Ceryle rudis* 25–26cm

A distinctive and unmistakable medium-sized kingfisher. Adult male has striking black and white upperparts. Underparts essentially white but has two distinct black breast bands. Adult female is similar to male but has one (not two) breast bands. Juvenile is similar to adult female but breast band is grey not black. Legs and large bill are black in all birds. Often perches beside open water but will also hover when vantage points are not available. Usually favours freshwater lakes and rivers but also occurs on sheltered coasts, especially outside the breeding season. Scarce resident in E Mediterranean, particularly along S coast of Turkey and in Middle East; occasionally wanders westwards in winter.

4. COMMON KINGFISHER *Alcedo atthis* 16–17cm

A stunning wetland bird. Has orange-red underparts and mainly blue upperparts; electric-blue back seen to best effect when bird is observed in low-level flight. Feet and legs are bright red. Sexes similar except that male has all-dark bill whereas that of female has reddish base to lower mandible. Invariably seen near water and uses overhanging branches to watch for fish. Plunges headlong into water to catch prey in bill; fish is sometimes stunned by beating head on branch before being swallowed head first. Nests in holes excavated in waterside bank. Flight call is a whistling *tsii*. Widespread resident breeding species from Iberia to Greece and, more locally, in NW Africa. In E Mediterranean, status is essentially that of a widespread winter visitor. Favours rivers and lakes but often forced to move to coasts if preferred sites dry up.

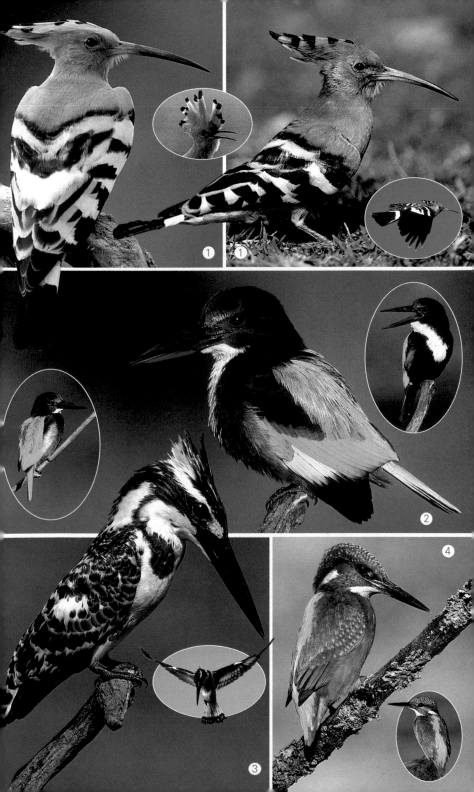

1. EURASIAN WRYNECK *Jynx torquilla* 16–17cm

An atypical woodpecker whose finely marked plumage is patterned like tree bark and affords the bird superb camouflage. Adult has marbled and streaked grey, black and brown upperparts. Shows dark eyestripe, which continues as a line down side of neck and across scapulars. Dark line runs from centre of the crown, down the neck to the back. Underparts are whitish with dark streaks and bars, and a yellow-buff wash on throat and upper breast. Juvenile is similar to adult but lacks the dark crown stripe. Feeds mainly on the ground, especially on ants. Nests in tree holes. Generally silent, but on breeding grounds call is a raptor-like piping *tiu-tiu-tiu...*. Favours open woodlands and orchards. Summer visitor to parts of Europe (May–Sept) and local resident or winter visitor in S Europe and N Africa; widespread passage migrant elsewhere.

2. BLACK WOODPECKER *Dryocopus martius* 42–45cm

Large and impressive woodpecker with essentially all-black plumage. Male has entire red crown while in female colour is restricted to rear half of crown. Eye and bill are pale in both sexes. In flight, wings are broad with splayed primary 'fingers'. Flight is direct but often looks laboured. Calls include a screaming *pree-pree-pree...* and a croaking *krruk-krruk*. In spring, male advertises territory with extremely loud drumming. Favours mature forest. Scattered resident, most common in centre and E of region.

3. EUROPEAN GREEN WOODPECKER *Picus viridis* 32–35cm

Medium-sized woodpecker with olive-green upperparts and buffish-grey underparts. Across most of region, adult male has a red crown, black around the pale eye, and a red-centred black moustachial stripe; female is similar but moustache is all black. Birds of Iberian subspecies *P. v. sharpei* are similar but both sexes lack the black around the eye. Juveniles (of all races) are similar to their respective adults but upperparts are marked with pale spots and underparts are adorned with numerous dark spots. In flight, all birds show a yellow-green patch on the rump and lower back. Flight is undulating. Call is a yelping *ku-ku-kyuk*. Favours open woodland and orchards; often feeds on the ground. Widespread resident along N shores of Mediterranean; absent from most islands.

4. GREY-HEADED WOODPECKER *Picus canus* 28–30cm

Resembles Green Woodpecker but is smaller. Upperparts are olive-green and underparts are olive-buff, grading to grey on the head and neck. Male has a red patch on the forecrown, black between the base of the bill and the eye, and a narrow black moustachial stripe. Female is similar to male but crown is entirely grey. Juvenile is similar to adult but colours are grubby. Call is a series of whistling notes, *ku-ku-ku-ku...*. Flight is undulating, revealing the yellow rump and lower back. Favours wooded parkland or inundated forest. Local resident from central France to N Greece, and northwards to Romania and beyond.

5. LEVAILLANT'S GREEN WOODPECKER *Picus vaillantii* 30–32cm

N African counterpart of Green and Grey-headed Woodpeckers and similar, in many respects, to both species. Upperparts are olive-green and underparts are olive-buff, grading to grey-buff on head and neck. Male has a red crown and nape, black between base of bill and pale eye, and a white-bordered black moustachial stripe; female is similar but crown is grey (red is confined to nape). Flight is undulating, revealing yellow rump and lower back. Call is a yelping scream, similar to that of Green Woodpecker. Favours upland forests. Resident in NW Africa.

1

2

3. *sharpei*

3

3. juv

5. ♀

5. ♂

4

1. GREAT SPOTTED WOODPECKER *Dendrocopus major* 23cm

The region's most common black and white woodpecker. Adult female has mainly black upperparts, with white patches and bars on the wings; black moustache connects with the black nape and the black bar on the shoulder, creating enclosed white cheeks. Underparts are grubby white except for red on the vent and undertail. Adult male is similar but has a small red patch on the nape. Birds of the N African subspecies *D. m. numidus* are similar to their European counterparts but show red on the breast and more extensive red on the belly. Juvenile is similar to adult but has a red cap. Flight is undulating, revealing striking white shoulder patches and rounded, chequered wings. Alarm call is a sharp *tchik*. In spring, male drums loudly. Nests in tree holes. Favours broadleaved woodlands. Widespread resident in Europe; local in Turkey and NW Africa.

2. LESSER SPOTTED WOODPECKER *Dendrocopus minor* 14–15cm

Tiny, unobtrusive woodpecker. Adult has barred, black and white upperparts, creating a ladder-backed appearance; underparts are whitish. Has white face with black nape, and black stripe from breast to cheek and base of bill; crown is red in male but black in female. Mainly silent but utters raptor-like call in spring. Nests in holes in branches. Favours broadleaved woodlands. Local resident.

3. MIDDLE SPOTTED WOODPECKER *Dendrocopus medius* 20–22cm

Smaller than Great Spotted and with a more delicate bill. Adult has a red crown, brightest and most extensive in male. Upperparts are mainly black but shows a bold white 'shoulder' patch and white barring across the flight feathers. Face is mainly white; black moustache does not connect with black stripe on nape. Breast and flanks have black streaks; underparts show buffish wash to belly grading to pink on vent. Favours open orchards and broadleaved woodlands. Excavates holes in decaying trees for nesting. Generally silent (does not drum) but sometimes utters subdued *kvah-kvah…* calls. Local resident, most widespread in E.

4. WHITE-BACKED WOODPECKER *Dendrocopus leucotos* 25cm

Larger than Great Spotted Woodpecker, but beware confusion with juvenile of that species. Note the extensive white barring on the black upperwings and the absence of a white 'shoulder' patch. Represented in the region mainly by subspecies *lilfordi*, adults of which have black and white bars (not pure white) on the back, heavily streaked underparts and variably barred flanks. Black stripe from moustache to upper breast sweeps back and just reaches black nape. Crown is red in male, black in female. Call and drumming are similar to that of Great Spotted. Favours mature forests with decaying trees. Local and scarce resident.

5. SYRIAN WOODPECKER *Dendrocopus syriacus* 23cm

Similar to Great Spotted, but black stripe that runs from moustache to side of breast does not connect with black on nape and is separated by a white gap. In addition, adults have pinkish red (not intense red) on the vent and variable streaking on the flanks. Note the white (not black) feathering at base of bill. Male has a red patch on nape while, in female, nape is all black. In all adult birds, pattern of white on wings is similar to that of Great Spotted. Juvenile is similar to adult but has a red crown and more extensive streaking on flanks and breast. Favours vineyards, gardens and orchards. Resident in E; range is expanding.

1

2

3

numidus   juv

4

5

1. COMMON HOUSE MARTIN *Delichon urbica* 12–13cm

Striking, small and pied hirundine. Identified in flight, in all plumages, by the conspicuous white rump on otherwise dark upperparts, and by the white underparts. In good light, upperparts of adult appear shiny bluish black; upperparts of juvenile are dull black. Feeds on the wing and often over water. Utters twittering, chirping calls at colonies. Nests colonially on buildings, constructing almost spherical mud nests under eaves and overhangs. Widespread summer visitor, and passage migrant throughout the region, present Mar–Sept.

2. SAND MARTIN *Riparia riparia* 12cm

Small hirundine, often seen in groups feeding over water. In flight, upperparts of adult are sandy-brown. Seen from below, shows white belly separated from white throat by brown chest band; white on throat extends up side of neck as half-collar. Tail is short and slightly forked. Juvenile is similar to adult but feathers on back have buffish margins. At colonies, birds utter harsh twittering calls. Nests in colonies in sand banks. Common passage migrant and widespread summer visitor, nesting from Iberia to E Mediterranean, present Mar–Sept.

3. EURASIAN CRAG MARTIN *Ptyonoprogne rupestris* 14cm

A bulky hirundine. Reminiscent of Sand Martin but is larger and has relatively broader wings. Upperparts of adult are uniformly dusky brown; has a dark chin and dark underwing coverts but underparts are otherwise uniform grey-buff (note the absence of a chest band, as seen in Sand Martin). In flight, shows diagnostic pale patches near tip of tail. Juvenile is similar to adult but feathers on back have buffish margins. During summer months, typically favours mountains and gorges, but in winter usually moves to lower elevations, where it often feeds over coastal marshes. Widespread resident.

4. BARN SWALLOW *Hirundo rustica* 19cm

Generally the most common and easily observed hirundine in the region. In silhouette, can be recognised by its pointed wings and long tail streamers; these are shorter in juvenile and female than male. Adult upperparts are glossy blue-black and, across most of the region, underparts are white except for the brick-red throat and forecrown, and the blue-black chest band; birds from the Middle East have white elements of underparts replaced by buffish-red. Juvenile is similar to adult (of corresponding race) but throat and forecrown are buffish pink, not red. Groups of birds often sit on wires. Utters *vit* call in flight; male has a twittering song. Nests in buildings. Mainly a widespread summer visitor and passage migrant, present Mar–Oct. Overwinters, in small numbers, in S Iberia and N Africa; resident population in Middle East.

5. RED-RUMPED SWALLOW *Hirundo daurica* 16–17cm

Similar to Barn Swallow but has bulkier body and tail streamers that curve inwards. Tail, vent and lower rump are black in all birds, with such a discrete cut-off that birds look as though the tail end has been dipped in black paint. Adult has shiny bluish-black upperparts except for the buffish-orange nape and pale rump. Underparts, including underwing and throat, are pale; shows streaks on breast. Juvenile is much duller than adult and has shorter tail streamers; rump is almost white. Call is a nasal *tchreek*. Makes cup-shaped mud nest in abandoned buildings and under bridges. Summer visitor and passage migrant, most common in Iberia, Greece and Turkey; present Apr–Sept.

1. CRESTED LARK *Galerida cristata* 17cm

Larger than Common Skylark and with longer bill and spiky crest. Upperparts are streaked sandy-brown; shows streaking on chest and flanks but underparts are otherwise pale. Note the pale supercilium, eye-ring and throat, and dark moustachial stripe. Outer tail feathers and underwing coverts are orange-buff. Favours dry habitats with sparse vegetation; often feeds beside roads and hence arguably the easiest lark species to observe in the region. Fluty song contains elements of mimicry. Widespread resident but absent from most islands in W of region.

2. DUPONT'S LARK *Chersophilus duponti* 17–18cm

An enigmatic species that is shy and difficult to observe. Body shape resembles that of Common Skylark (or perhaps a short-tailed pipit), but note the long, downcurved bill. Neck is often extended in nervous birds, adding to pipit-like appearance. Upperparts are grey-brown (Iberia and NW Africa) or sandy-brown (elsewhere in range); neck is streaked but underparts are otherwise pale. Legs are relatively short and stout. Wings appear uniformly brown in flight. Fluty, whistling song is difficult to pinpoint. Favours steppe-type habitats with sparse clumps of grass and other vegetation. Despite the open appearance of its favoured terrain, is extremely hard to find; typically hides behind cover and runs (rather than flies) from danger. Resident in Iberia and N Africa.

3. THEKLA LARK *Galerida theklae* 17cm

Similar to Crested Lark but distinguishable with care. Note Thekla Lark's shorter, thicker bill (lower mandible is convex) and more complete crest. Adult has streaked grey-brown upperparts. Streaks on breast and flanks are more distinct than on Crested (almost thrush-like spots) but underparts are otherwise pale. Underwings are pale greyish buff. Song is similar to that of Crested Lark. Often perches on bushes. Favours garrigue and steppe habitats. Widespread resident in Iberia and NW Africa; occurs on Balearic Islands (where Crested Lark is absent) and locally in S France.

4. WOOD LARK *Lullula arborea* 15cm

Small and rather boldy marked lark with a short tail. Adult has streaked sandy-brown upperparts; underparts are white except for streaking on breast. Note the dark-bordered chestnut ear coverts and white supercilia that meet on nape. Wings have a black and white patch at the bend of leading edge, visible both in flight and at rest. Juvenile is similar to adult but is less well marked. Delightful yodelling song is usually delivered in flight; call has quality reminiscent of the song. Favours dry terrain with short vegetation and scattered trees; also found in open pine woodland. Mainly a widespread but local resident and winter visitor to N shores of Mediterranean, NW Africa and Middle East.

5. COMMON SKYLARK *Alauda arvensis* 18cm

A rather nondescript lark with a stout bill and short erectile crest. Upperparts are sandy-brown and streaked. Has streaking on the breast but underparts are otherwise whitish. Note the pale supercilium and buffish cheeks. Wings show pale wingbars at rest and a white trailing edge in flight; tail has white margins. Incessant trilling song is often sung in flight; resident birds often sing year-round. Calls contain elements of song. Favours grassy areas and stubble fields. Occurs year-round along N shores of Mediterranean and NW Africa; numbers boosted outside breeding season by influx of birds from N Europe. Mainly a winter visitor to Middle East and NE Africa.

1. GREATER SHORT-TOED LARK *Calandrella brachydactyla* 13–15cm

Small, dumpy lark with a rather short, finch-like bill. Sandy-brown upperparts have faint streaks. Note the broad, pale supercilium, brown cheeks and white throat; streaked crown is sometimes tinged rufous. Underparts are mainly white with dark patches on sides of breast (these can be hard to see at some angles); some birds show faint streaking across chest. Median coverts are dark-centred with pale margins. Long tertials overlap primaries and nearly reach tip of wing in standing bird. Call is a sharp *trilp*. Trilling song often contains elements of mimicry. Favours dry, grassy places and arable fields. Typically a widespread passage migrant and summer visitor, present Apr–Sept. During migration, the most numerous lark in the region, with flocks – often comprising hundreds of birds – found on coastal fields. Seemingly resident populations also occur locally.

2. LESSER SHORT-TOED LARK *Calandrella rufescens* 13–15cm

Similar to Greater Short-toed Lark, but can be distinguished with care by noting differences in plumage, call and habitat. Note the short, stubby bill, the rounded head, and the streaked breast band (and absence of dark patch on side of breast found in Greater Short-toed Lark). A close view reveals the shorter tertials (primaries are exposed in standing birds). Upperparts are streaked and grey-brown, while underparts (apart from streaked breast) are whitish. Median coverts are dark-centred with pale margins but contrast is not so striking as in Greater Short-toed Lark. Call is a buzzing *tchrrrt*. Song includes elements of mimicry. Favours dry, saline, lake margins and sparse steppe. Locally common resident in Iberia, N Africa and the Middle East; passage migrant or local winter visitor elsewhere in E of region.

3. BIMACULATED LARK *Melanocorypha bimaculata* 17–18cm

A relatively large and large-billed lark. Similar to Calandra Lark but distinguishable with care. Note the even heavier bill and more distinct and contrasting appearance to the plumage (particularly the face). Upperparts are streaked grey-brown and face is marked with a pale supercilium and throat, and a dark line through and below the eye. Underparts are white except for a narrow dark patch and streaks on the sides of the breast. In flight, note the uniform grey-brown underwings and the white-tipped tail. Call is a sharp *drrrup* and song is similar to that of Calandra Lark. Favours steppe and semi-desert habitats. Mainly a summer visitor (Apr–Sept) to E of region (from E Turkey eastwards) but typically a passage migrant or winter visitor to the Middle East.

4. CALANDRA LARK *Melanocorypha calandra* 18–19cm

Large and impressive lark with a relatively large head and bill. Tail is relatively short compared to other lark species. Upperparts are sandy-brown with dark streaking on the crown, nape and back. Face is marked with a pale supercilium, eye-ring and throat. Underparts are mainly white, but note the bold black patches on the sides of the breast and the small spots that extend down the flanks. Tail has white outer feathers. In flight, shows dark trailing margin to wing and diagnostic black underwing; latter feature is easiest to see when bird is singing. Song is full-bodied and trilling, with elements of mimicry. Call is a harsh *schreek*. Favours steppe habitats and areas of short grassland (even arable and stubble fields). Typically a widespread resident, generally most numerous in Iberia, NW Africa, Greece and Turkey. Outside breeding season (when numbers boosted by migrants from further N), forms flocks and is rather nomadic.

1. BAR-TAILED LARK *Ammomanes cincturus* 13–14cm

Uniformly plain lark. Superficially similar to Desert Lark but distinguishable with care. Upperparts are uniformly sandy-brown, the back unmarked and the wings showing pale margins to the feathers; in resting bird, tertials appear tinged reddish and primary tips are black. Underparts are whitish with little or no streaking on the breast. Tail is relatively short and reddish brown with a narrow and well-defined black terminal band. Head is rather rounded. Bill is shorter and stubbier than Desert Lark and uniformly pinkish yellow. Call is a soft, trilling *chrrt*. Subdued, thin song is often delivered in flight. Favours flat desert and semi-desert areas. Resident in N Africa and the Middle East.

2. DESERT LARK *Ammomanes deserti* 15–16cm

Plain-looking lark with a compact body, a relatively long tail and a robust, triangular bill. Upperparts are uniformly sandy-brown; the back is unmarked while the wings show pale margins to the feathers and a reddish tinge to the primaries. In some races, upperparts are extremely pale. Underparts are whitish buff with faint streaks on the breast. Tail is reddish brown with a broad and diffuse black terminal band. Bill is pinkish yellow at the base but with a dark tip and culmen. Call is a soft *chuu*. Ringing song is often delivered in flight. Favours stony desert slopes and wadis. Resident in N Africa and Middle East.

3. HORNED LARK (SHORE LARK) *Eremophila alpestris* 15–17cm

Dumpy and distinctive lark. At all times, adults have grey-brown or sandy-brown upperparts that show faint streaking, and whitish underparts. Head pattern is always characteristic, with its yellow and black markings. In summer, adult male has long black 'horns' projecting from sides of crown; in female and winter birds of both sexes, the 'horns' are reduced or absent and the yellow on the face is less intense. Juvenile resembles winter adult but has scaly and spotted appearance. Call is a thin *psee-pseea*. Breeds on stony mountains above tree-line, but sometimes descends to similar-looking habitats at lower altitudes in winter. Local resident in NW Africa, from Balkans eastwards, and in Middle East.

4. TEMMINCK'S LARK *Eremophila bilopha* 13–15cm

Resembles a small Horned Lark with washed-out plumage. Upperparts of adult are mainly greyish sandy-brown; in resting birds, note the reddish wing coverts and the rufous-tinged tertials, contrasting with the blackish primary projections. Underparts are whitish and unmarked. Head pattern is distinctive, with black and white markings and black 'horns' projecting from sides of crown. Juvenile resembles adult but upperparts are scaly reddish brown and adult's head pattern (including 'horns') is absent. Call is a thin *psee-pseea*. Favours stony desert slopes and semi-deserts. Resident in N Africa and the Middle East.

5. GREATER HOOPOE LARK *Alaemon alaudipes* 20–22cm

Comparatively large and readily identifiable pipit-like lark. Note in particular the long, slender and slightly downcurved bill, and the relatively long legs and tail. Upperparts are mainly uniformly pale sandy-brown to greyish brown. At rest, the wing feathers appear dark-centred and pale-fringed; in flight, wings reveal a striking and diagnosic black and white pattern. Head shows a dark eyestripe and moustachial stripe on the otherwise pale face. Underparts are pale except for variable streaking on the breast. Runs at speed in preference to flying. Has a buzzing call and evocative, ringing song. Favours deserts and semi-deserts. Resident in N Africa and the Middle East.

1. TREE PIPIT *Anthus trivialis* 14–15cm

Similar to Meadow Pipit but distinguishable with care using subtle plumage differences, habits and voice. Upperparts are olive-brown and generally more heavily streaked than on Meadow Pipit. Underparts are pale and boldly streaked, showing contrast between the white belly and buff-flushed breast. Note the more pronounced pale eye-ring and marginally larger bill. Call is buzzing *spzzzt*. Trilling song is delivered in parachuting flight, usually starting from treetop perch. Favours open grassy places on migration but open woodland for breeding. Widespread passage migrant and breeds from N Iberia to N Turkey northwards; present Apr–Sept.

2. MEADOW PIPIT *Anthus pratensis* 14–15cm

The archetypal pipit. Has streaked brown upperparts and paler underparts, the latter with a buffish wash to the breast and streaks on the throat, breast and flanks. Shows two pale wingbars and a pale supercilium. The legs are pinkish-flesh. White outer tail feathers are easiest to see in flight when bird utters *pseet-pseet-pseet* call. In early spring, migrants sometimes deliver trilling, descending song in flight. Favours short grassland. Forms flocks. Mainly a common winter visitor (Oct–Mar) but present year-round locally in NW.

3. RED-THROATED PIPIT *Anthus cervinus* 15cm

Boldly marked pipit. In spring, adult has brick-red face and neck. Intensity and extent of colour varies (and males are generally more strongly marked than females), and birds with buffish faces are often seen. Underparts are otherwise pale buff to whitish, streaked heavily on chest and flanks. Upperparts are dark brown, heavily streaked with black and white. Autumn birds appear essentially blackish-brown and white with heavy streaking on breast. Legs of all birds are buffish yellow. Call is a thin *tseeeh*. Favours short grassland and often feeds in flocks. Passage migrant, commonest in E; winters locally in N Africa and Middle East.

4. TAWNY PIPIT *Anthus campestris* 16–18cm

A pale, long-legged pipit. Adult upperparts are rather uniformly sandy-brown and only faintly streaked. Wings show wingbars formed by pale-fringed, dark-centred covert feathers. Face is pale with a long, pale supercilium. Underparts are pale and mostly unmarked. Tail is proportionately long and brown with pale edges to outer feathers. Juvenile resembles adult but upperparts and breast are more heavily streaked. Legs are long and flesh-coloured in all birds. Utters House Sparrow-like *chirrupp* call; repetitive song comprises bursts of short call-like phrases. Favours dry sandy habitats with scattered grass. Mainly a widespread summer visitor (Apr–Sept) but winters locally in Middle East.

5. WATER PIPIT *Anthus spinoletta* 16–17cm

Robust and dumpy pipit with variable plumage. Summer adult has a grey head with a white supercilium and throat, a pinkish flush to the breast on otherwise white, unmarked underparts, and a brown back and tail. Winter adult has brown, faintly streaked upperparts and pale buffish-white underparts, boldly marked with black streaks. At all times, legs are reddish brown and tail shows white outer feathers. Call is a drawn-out *tseeest*. Breeds on mountains but winters on coastal and low-lying wetlands. Widespread but local resident and winter visitor. Similar **Buff-bellied Pipit** *A. rubescens* (not illustrated) is a scarce visitor to Middle East. Recalls an outsized Meadow Pipit, boldly marked below but mainly unstreaked olive-grey above.

2. autumn

3

imm

3. winter

1

2

3

4

5

1. GREY WAGTAIL *Motacilla cinerea* 18cm

Attractive long-tailed, slim-bodied bird with blue-grey upperparts and lemon-yellow breast and belly. In summer, adult male has a black bib-like throat bordered above by a white moustachial stripe; also shows dark eyestripe with white supercilium. In winter male (and females at all times), the pure black throat becomes mottled or is replaced by white. Always associated with water, mostly fast-flowing streams and rivers. Perches on boulders, pumping tail up and down. Utters *chsee-tsit* call in bounding flight. Widespread but local resident although essentially a winter visitor to E of region and to N Africa.

2. WHITE WAGTAIL *Motacilla alba* 18cm

Long-tailed, slim-bodied black and white bird. Often seeing running on ground, typically pumping tail up and down. Adult male in summer has black cap, nape, throat and upper breast. Face and underparts are white while back is grey. Tail has white outer feathers. Winter male has black confined to upper breast only; plumage otherwise similar to summer male but grubbier. Summer female is similar to summer male but contrast is less intense; winter female and juvenile are similar but black cap is lost and plumage appears grubby. Call is a sharp *chissick*. Favours bare, grassy areas. Widespread resident along N shores of Mediterranean and NW Africa. Numbers boosted outside breeding season by migrants from N and winter range extends across N Africa to Middle East.

3. YELLOW WAGTAIL *Motacilla flava* 16–17cm

Slim-bodied bird. Comprises distinct geographical races; six of the most regularly occurring are described here and males are distinguishable using head markings. All adult males have greenish-yellow mantles, darker wings with two pale wingbars, and a dark tail with white outer feathers; underparts are bright yellow. **Blue-headed Wagtail** *M. f. flava* has bluish cap and cheeks, a white supercilium, a white chin and a yellow throat; **Spanish Wagtail** *M. f. iberiae* is similar but cap is greyer, cheeks are darker and throat is white; **Grey-headed Wagtail** *M. f. thunbergi* has grey head, dark cheeks, a yellow throat and no supercilium; **Ashy-headed Wagtail** *M. f. cinereocapilla* has dark grey cap, black cheeks, white throat and no supercilium. Adult male Yellow Wagtail *M. f. flavissima*, and females of all races, have yellow heads. **Black-headed Wagtail** *M. f. feldegg* is sometimes considered to be a separate species. Adult male has wholly black cap and cheeks, a yellow throat and underparts, and an olive-brown back; adult female has greyer upperparts and whiter underparts than females of other races. Other subspecies occur in E of region, for example *M. f. dombrowskii*. Typical call is a thin *tsr-ree*; that of *M. f. feldegg* is more grating than with other races. Favours marshes and damp grassland. Species is typically a widespread passage migrant and summer visitor (Apr–Sept). Race *feldegg* is regular only in E. N African populations are resident and small numbers winter in S Spain.

4. CITRINE WAGTAIL *Motacilla citreola* 16–17cm

Distinctive, slim-bodied bird. All birds have a grey back, white-fringed dark-centred wing feathers (note the white wingbars), and white undertail coverts. Adult male in summer has rich lemon-yellow on the head and underparts, and a black nape and half-collar. Adult female has a pale yellowish wash to the supercilium, throat and underparts, and a darker cap and cheeks. First-winter birds have essentially grey upperparts and whitish underparts. Note, however, the striking white wingbars and the pale border to the darker ear coverts (the pale supercilium is linked to the pale throat). Call is a shrill *tsreep*. Favours water margins on migration. Regular passage migrant in E; scarce in W.

① winter

juv

② ♂

summer ♂

♂

thunbergi ♂

2. winter

flavissima ♂

flavissima ♀

dombrowskii ♂

iberiae ♂

flava ♂

feldegg ♀

cinereocapilla ♂

③

feldegg ♂ ③

④ ♂

juv

1. WINTER WREN *Troglodytes troglodytes* 9–10cm

One of the region's smallest birds. Recognised by its brown plumage, dumpy proportions and habit of cocking its tail upright. Upperparts are rufous-brown with faint barring on the back, and underparts are paler buffish brown and barred. Closed wing shows barring on primary feathers and head has prominent pale supercilium. Song is loud and warbling, ending in a trill; has a rattling alarm call. Favours dense cover in scrub and woodland. Widespread resident from Iberia to Turkey northwards; local in N Africa and Middle East.

2. WHITE-THROATED DIPPER *Cinclus cinclus* 18–20cm

Dumpy, essentially black and white bird. Invariably seen perched on boulders in fast-flowing rivers. Adult upperparts are dark brown, most rufous on the head; underparts comprise a white 'bib' (throat, neck and upper breast), clearly separated from the darker belly (chestnut at the front grading to blackish at the rear). Juvenile is grey and scaly, palest on the underparts. Call is a shrill *stritts*. Typically bobs body up and down when perched; feeds underwater. Widespread but local resident; absent from much of N Africa and Middle East.

3. ALPINE ACCENTOR *Prunella collaris* 18cm

Adult has dark brown upperparts and a rufous rump; the head, breast and belly are blue-grey but otherwise the underparts are pale with chestnut streaks on flanks and black and white undertail coverts. Closed wing is marked with white dots and the white chin is speckled with black dots. Juvenile has subdued colours compared to adult. Breeds in mountains; often descends to lower rocky outcrops and coastal headlands in winter and present there Oct–Mar.

4. DUNNOCK *Prunella modularis* 14–15cm

Plump-bodied and unobtrusive bird with a needle-like bill. Adult has a chestnut-brown back marked with dark streaks, blue-grey underparts with a rufous wash to the streaked flanks, and brownish cheeks and crown. Juvenile is similar but plumage is more uniformly brown and streaked. Feeds quietly, often on ground, searching for insects and seeds. Call is a thin *tseer*. Resident breeding species in inland (including upland) areas; numbers boosted and range extended S by winter visitors (Oct–Mar).

5. EUROPEAN ROBIN *Erithacus rubecula* 14cm

Distinctive, dumpy bird. Adult has reddish-orange face and breast with a whitish belly and grey-brown flanks; upperparts are brown, grading to grey on the side of the neck. Juvenile has streaked brown upperparts and underparts with crescent-shaped markings. Call is a sharp *tic*. Male sings melancholy song at most times of year. Favours woodland and scrub. Widespread resident from Iberia to N Turkey northwards, and in NW Africa. More numerous and widespread in winter.

6. BLUETHROAT *Luscinia svecica* 13–14cm

Attractive bird with Robin-like proportions. All birds have grey-brown upperparts, a pale supercilium, whitish belly and undertail coverts, and a dark-tipped tail that is rufous at the base. Summer males have a blue throat, defined below by bands of black, white and red; different races show either a white or a red central throat spot, or an unspotted throat. Adult females are variable, some with white throats (like first-winter birds), others with a hint of blue on an otherwise pale throat (like winter adult males). Call is sharp *tchak*. Favours wetlands. Mainly a passage migrant and local winter visitor (Oct–Mar); local breeding also occurs.

① ② ③ ④ ⑤

juv

1st-w

ad w ♂

♂

⑥

1. COMMON NIGHTINGALE *Luscinia megarhynchos* 16–17cm

Unobtrusive bird, best known for its wonderful song. Heard far more easily than the bird itself is seen. Has rich brown upperparts and a chestnut-red lower back and tail; underparts are pale grey-buff. Note also the pale eye-ring and faint, uniformly buff breast band. Song is loud and musical, and is delivered both by day and at night. Favours woodland with a dense scrub layer. Widespread and common summer visitor and passage migrant; present Apr–Aug.

2. THRUSH NIGHTINGALE *Luscinia luscinia* 16–17cm

Very similar to Nightingale; can be impossible to separate if views are brief or partial, and if bird is silent. Note the grey-brown upperparts and rufous-tinged tail. Pale underparts show a grey-brown breast band, grading to faint thrush-like spots on the flanks (most noticeable in first-winter birds). Faint grey-brown moustachial stripes and shorter first primary than that of Nightingale are not useful features in the field. Song is slower, deeper and less complex than Common Nightingale, and often includes throaty clucking sounds. Favours damp woodland and scrub. Passage migrant through E of region; local breeder in NE.

3. BLACK REDSTART *Phoenicurus ochruros* 14cm

Striking red tail is seen in all plumages. Breeding male is otherwise blackish, with orange vent and undertail coverts; European males have a white wing patch while Middle East males have an orange-red belly. Female (and winter male) is grey-brown. Song is whistling and repetitive; call is a thin *tsiist*. Favours rocky slopes and villages. Occurs year-round in N and W (winter numbers boosted by migrants). Mainly a winter visitor to E, and to N Africa.

4. COMMON REDSTART *Phoenicurus phoenicurus* 14cm

All birds have a dark-centred orange-red tail. Adult male has grey on back, nape and crown, a black face and throat, and a white supercilium; breast is orange-red, grading to whitish on vent (E males also have a white wing patch). Female has grey-brown upperparts and an orange wash to pale underparts. First-winter male resembles female but has hint of head pattern seen in adult male. Song is musical and whistling; call is a warbler-like *huiit*. Favours open broadleaved woodland. Passage migrant and local breeding species, present Apr–Sept.

5. MOUSSIER'S REDSTART *Phoenicurus moussieri* 12–13cm

Small, dumpy redstart. Male has orange-red underparts and black upperparts, with a white wing patch and broad white supercilium that extends down side of neck. Female has brown upperparts and orange-buff underparts. All birds have an orange-red rump and tail. Song is a whistling warble. Favours scrub-covered slopes. Resident in NW Africa.

6. WHITE-THROATED ROBIN *Irania gutturalis* 17–18cm

Unmistakable male has a blue-grey back and wings, and a black tail; head shows a blue-grey crown and nape, a white supercilium, and a black face with a white throat stripe. Breast and belly are orange-red, grading to white on the vent. Adult female and first-winter bird are mostly grey with an orange flush on the flanks, a white vent and a pale throat and moustachial stripes; some birds show orange-buff crescent markings on the breast. Male sings a warbling song at start of breeding season, often from conspicuous perch. Otherwise, all birds are generally skulking. Favours areas of low scrub with scattered trees. Passage migrant in E of region and breeds regularly from Turkey eastwards; present Apr–Aug.

Middle East ♂

Eastern ♂

1. BLACK-EARED WHEATEAR *Oenanthe hispanica* 14.5cm

Plumage is variable, but all birds have a white rump and tail marked with an inverted black 'T'. Spring males occur in two morphs, either with a black throat or a black mask through the eye; both have black wings. W males have sandy-buff on the crown, nape and mantle, and pale underparts, suffused buff on breast; in E males, these elements of plumage are pale, sometimes almost white. Females of both races are similar to their respective males but with subdued plumages. Call is a grating *grssch*; song is scratchy. Favours broken, stony ground. Widespread passage migrant and summer visitor, present Apr–Sept.

2. ISABELLINE WHEATEAR *Oenanthe isabellina* 16.5cm

Similar to female or first-winter Northern Wheatear. Appears marginally larger than that species, generally adopts a more upright stance on the ground, and appears to have relatively long legs. All birds have rather uniformly grubby sandy-buff upperparts; underparts are pale but suffused with buff on the throat and breast. Flight feathers have pale fringes and alula appears contrasting dark against the otherwise pale wing. All birds reveal a white rump in flight. Call is sharp *tchiut* and song includes wader-like whistles and mimicry. Favours dry, sandy terrain and steppe habitat. Summer visitor and passage migrant to E of region (Apr–Sept); small numbers winter in Middle East.

3. NORTHERN WHEATEAR *Oenanthe oenanthe* 14–15cm

Familiar chat of open country, all birds of which show a white rump in flight. Adult male in spring has a blue-grey crown and back, a black mask bordered above and below by white, black wings, and pale underparts with an orange-buff wash on the breast (NW African birds also have a black throat). Female resembles male but black mask is absent, wings are brown (not black) and back is brownish grey. In autumn, all birds look more sandy-brown with pale-fringed dark wing feathers; first-winter birds are palest (particularly on the head). Call is a sharp *tchack*, like two pebbles being knocked together; song is scratchy, with harsh phrases. Favours open grassland, heaths and cliffs. Passage migrant and widespread summer visitor, present mainly Mar–Oct.

4. COMMON STONECHAT *Saxicola torquata* 12–13cm

Distinctive and compact chat with a relatively short tail. In breeding season, male has a black head, with white on the side of the neck, a dark back, an orange-red breast and pale underparts. Female and winter male have brown, not black on the head, and overall more subdued colours; throat is brown in summer female but generally pale in winter birds of both sexes. Call is a harsh *tchak*, like two stones being knocked together. Scratchy song recalls that of Common Whitethroat; delivered in flight or from song perch. Favours heaths, maquis and scrub patches. Widespread resident in Europe and NW Africa; numbers increase in winter months, when range extends eastwards to the Middle East.

5. WHINCHAT *Saxicola rubetra* 12–13cm

Similar proportions to Stonechat but male has brown, streaked upperparts and striking pale supercilium; underparts are pale with orange-buff flush on throat and upper breast. Female and first-winter birds (of both sexes) resemble male in plumage pattern (note the pale supercilium) but colours are generally much more subdued. Call is a sharp *tik* and song is chattering. Favours rank grassland and scrubby slopes. Typically perches on low wires and bushes. Widespread passage migrant and local breeding species, present Apr–Sept.

Western ♂

Eastern ♂

1

Eastern ♀

Eastern ♀

juv

3

1. Eastern ♂

3

2

3

3

♂

N African ♂

4

4. 1st-w

4

♀

♂

♂

5

1. PIED WHEATEAR *Oenanthe pleschanka* 15–16cm

Striking and boldly marked wheatear. Adult male has a black face and throat, continuous with the black back and wings. Crown, nape and underparts are whitish, suffused pale buff on the breast and greyish on the centre of the crown. Female has greyish-brown head, neck and back, with darker wings and pale underparts. First-winter birds resemble adult female but underparts are suffused with buff and upperparts are scaly owing to pale feather margins. In flight, all birds show a white rump and white tail marked with an inverted black 'T' (black also extends back a short distance along sides of tail). Call is a harsh *brssch*; song is scratchy. Favours barren stony slopes. Passage migrant through E of region; breeds locally from Romania and Bulgaria eastwards, present Apr–Aug.

2. FINSCH'S WHEATEAR *Oenanthe finschii* 15–16cm

Recalls both Pied and Mourning Wheatears. Adult male has a black face and neck linked to the black wings; breast and underparts are whitish buff, as are crown and nape (the pale extends and narrows down the upper back). Female has greyish upperparts (except for darker wings and brownish ear coverts) and pale underparts (throat is dark in autumn birds). In flight, all birds reveal a white rump and white tail marked with a short inverted black 'T'. Calls include a sharp *tchak-tchak*; song is varied and scratchy. Favours stony slopes. Local summer visitor to uplands from Turkey eastwards (Apr–Sept); winters from lowland Turkey eastwards, S to Middle East. Scarce passage migrant elsewhere in E.

3. CYPRUS WHEATEAR *Oenanthe cypriaca* 14cm

Similar to Pied Wheatear. Adult male has a dusky-centred white crown and nape, white underparts (flushed orange-buff on the breast) and otherwise black plumage. Female has similar plumage pattern to male but pale elements are suffused with buffish orange on underparts and grey-buff on crown. First-winter bird is similar to adult female. In flight, all birds show a white rump and white tail, marked with an inverted black 'T'. Call and song include unusual buzzing sounds. Favours stony ground. Summer visitor to Cyprus, present Mar–Sept. Scarce passage migrant elsewhere in E of region.

4. MOURNING WHEATEAR *Oenanthe lugens* 15–16cm

Most similar to Pied Wheatear. Adult male has a black face, upper neck, back and wings; underparts, crown and nape are white (white on nape shows a distinct cut-off from black back and does not taper as in Finsch's). Note that males from Middle East have reddish-buff undertail coverts; in this race, female has similar plumage to male. Female from N African race has grey upperparts (except for darker wings) and pale underparts (throat is dark in autumn birds). In flight, all birds reveal white rump and white tail marked with a short inverted black 'T'. Calls include a harsh *tschak*. Favours stony slopes and mountains. Resident in NW Africa and the Middle East.

5. HOODED WHEATEAR *Oenanthe monacha* 16–17cm

Well-marked wheatear. Adult male has black on the face, neck, upper breast, back and wings; crown and underparts are white. Adult female has grey-brown upperparts, with darker wings, and pale orange-buff underparts. In flight, adult male reveals a white rump and a white tail with central black feathers; pattern is similar in adult female but white element of plumage is pale buff. Calls include a harsh *tsschak* and thin whistles; song includes elements of calls. Favours upland stony slopes. Resident in the Middle East.

1. RED-RUMPED WHEATEAR *Oenanthe moesta* 15–16cm

Adult male has a black face and throat connected to the black back and wings (note white margins to wing coverts), and greyish white on the crown, nape, breast and belly; vent, undertail, rump and upper tail are reddish-buff while latter half of tail is black. Adult female has reddish-buff upperparts and paler underparts; colours and markings on rump and tail are similar to those of male. First-winter birds are similar to adult female but back is greyer. Utters a whistling call. Favours semi-deserts. Local resident in the Middle East and N Africa.

2. BLACK WHEATEAR *Oenanthe leucura* 16–18cm

Relatively large wheatear. Adult male has mainly black plumage; the vent and undertail coverts are white. Rump is white and tail is white but marked with an inverted black 'T'. Adult female and first-winter birds are similar to adult male but black elements of plumage are dark sooty-brown. Call is a thin *peeup*; song is warbling and thrush-like. Favours rocky outcrops and slopes. Locally common resident in Iberia and NW Africa.

3. RED-TAILED WHEATEAR *Oenanthe xanthoprymna* 14–15cm

Adult males from W of breeding range have black on face and throat, connected by black to the dark wings; upperparts are otherwise grey (note the pale supercilium) while underparts are pale, grading to reddish buff on the vent and undertail. In flight, note the red rump and white-sided tail marked with an inverted black 'T'. W females and first-winter birds have mainly grey-buff plumage, a reddsh rump and (usually) a reddish tail marked with an inverted black 'T'. Birds of both sexes from E of range are similar to female from W of range but grey on face is more apparent. Calls include whistles and a harsh *tchak*. Favours barren slopes. Mainly a scarce passage migrant in E. Rare summer visitor from E Turkey eastwards; (Apr–Sept); rare in winter in Middle East.

4. WHITE-CROWNED (BLACK) WHEATEAR *Oenanthe leucopyga* 17–18cm

A large, plump wheatear. Adults of both sexes have mainly black plumage, except for the striking white crown and white vent and undertail coverts. In flight, note the white rump and white tail with black central feathers. First-winter birds have an all-dark head and variable amounts of black on tips of outer tail feathers. Has a whistling call; tuneful song contains elements of mimicry. Favours deserts. Resident in the Middle East and N Africa.

5. DESERT WHEATEAR *Oenanthe deserti* 15cm

Adult male has black face and throat, linked to black wings; crown, nape and back are sandy-brown (note the pale supercilium). Pale underparts are suffused warm buff on breast. Female is similar but has sandy-buff upperparts and pale underparts, including throat. First-winter male resembles female but shows hint of adult male's dark face pattern. In flight, all birds reveal a white rump and all-dark tail. Call is a soft *che-ah*. Favours semi-deserts. Resident and partial migrant in N Africa and Middle East.

6. BLACKSTART *Cercomela melanura* 15–16cm

Has mainly blue-grey body plumage (darkest on the wings and palest on belly and undertail) and an all-dark tail. Bill, eyes and legs are black. Utters a buzzing call and whistling song. Typically spreads wings and fans tail on alighting. Favours stony deserts. Resident in Middle East.

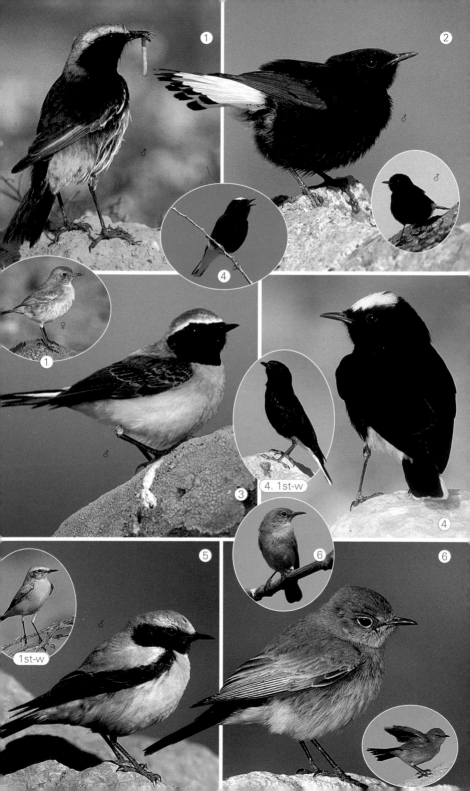

1. RING OUZEL *Turdus torquatus* 24cm

Similar to Blackbird in size and proportions. Adult male is mainly black with a striking white crescent on the breast; at close range, note the pale feather margins on the belly, and on the wing where they form a pale patch. Female is brown with conspicuous pale feather margins giving a scaly appearance; pale crescent on the breast is less striking than on male. First-winter bird is similar to female but crescent mark is subdued. Favours rough, stony slopes. Generally shy. Call is a sharp *tchuk*; song is whistling and fluty. Mainly a passage migrant and local winter visitor (Oct–Mar). Also a summer breeding visitor to some upland areas (these birds are typically altitudinal, not long-distance, migrants).

2. COMMON BLACKBIRD *Turdus merula* 25cm

Widespread and familiar woodland bird. Adult male has all-black plumage, with a yellow eye-ring and bill. Adult female has brown plumage, rather uniform above and streaked below; juvenile is similar to adult female but has numerous pale buff spots. Call is a harsh *tchak*, uttered at dusk and in alarm. Male's song is rich and fluty. Usually feeds on ground but perches regularly in trees. Widespread resident throughout Europe and NW Africa; numbers are boosted in winter by influx of birds from N Europe.

3. FIELDFARE *Turdus pilaris* 25–26cm

A large and well-marked thrush. Recognised by its blue-grey head and chestnut back; pale underparts are marked with dark spots, streaks and crescents, and are suffused yellow on the breast. Note also the pale supercilium, the pale grey rump and the white underwings, which are noticeable in flight. Calls include a chattering *tchak-tchak-tchak*. Usually seen in flighty flocks, sometimes mixed with Redwings. Essentially a winter visitor to the region, present Nov–Mar. Numbers and precise range are influenced by the severity of the winter further N in Europe. Small numbers breed in uplands in N of region.

4. REDWING *Turdus iliacus* 21cm

Recognised by its grey-brown upperparts, prominent white supercilium, and its pale underparts that are marked with dark streaks and flushed reddish orange on the flanks and underwings. Call is a high-pitched *tseerp*. Usually found in large, nomadic flocks and often associates with Fieldfares. Widespread winter visitor to Europe, Turkey, NW Africa and Middle East, present Nov–Mar. Numbers and range are influenced by winter weather further N in Europe.

5. MISTLE THRUSH *Turdus viscivorus* 27cm

Large, plump thrush. Upperparts are grey-brown and underparts are pale with large, dark spots. Note the faint white wingbars and, in flight, the white underwing coverts and the white tips to the outer tail feathers. Juvenile has white, teardrop-shaped spots on back and scaly underparts. Has loud, rattling alarm call; song contains brief phrases and long pauses. Favours open woodland. Generally solitary. Widespread resident and winter visitor.

6. SONG THRUSH *Turdus philomelos* 23cm

Small, well-marked thrush. Upperparts are warm brown and underparts are pale with dark spots and a buff wash to breast. Note the faint orange-buff wingbar and orange-buff underwing coverts. Song is musical and loud, the phrases repeated two or three times. Call is a thin *tik*. Favours open woods. Widespread winter visitor (Oct–Mar) and local resident breeder in inland upland regions.

1. RUFOUS BUSH ROBIN *Cercotrichas galactotes* 15cm

In terms of body appearance, resembles a cross between a chat and a large warbler; note, however, the broad, graduated tail. Birds from W of region (Iberia and NW Africa) have rufous-brown upperparts, brightest and most rufous on the rump and tail. Note the black and white markings at the tip of the tail (striking in flight and when tail is fanned), the pale supercilium and the dark eyestripe. Birds from E of region have mainly grey-brown, not rufous, body plumage; note, however, the rufous rump and lower back, and the rufous tail, tipped with black and white. Calls include buzzing and clicking sounds; song is thrush-like. Favours scrub areas; often in twiggy bushes that fringe drying rivers. Summer visitor to S Iberia, N Africa, Greece, Turkey and the Middle East, present Apr–Sept.

2. ROCK THRUSH *Monticola saxatilis* 17–19cm

Colourful and distinctive thrush-like bird. Summer male has a blue head, neck and upper back, dark brown wings and orange-red underparts and tail; the white lower back is most readily seen in flight. Adult female and juvenile have scaly brown upperparts, brown wings with pale feather margins, and orange-buff, scaly underparts; tail is red. Winter male is similar to female but head is greyer. Calls include a sharp *tchak*; song is fluty and melodic. Often perches upright, in the manner of a chat or wheatear. Generally rather wary and quick to hide or fly from danger. Favours dry rocky slopes. Widespread passage migrant and locally common summer visitor, present Apr–Sept.

3. BLUE ROCK THRUSH *Monticola solitarius* 20–22cm

Thrush-like bird with a proportionately long bill. In poor light, can appear all dark. Adult male has mainly dull blue plumage, brightest on head and underparts; wings and tail appear contrastingly darker and brown. Female and juvenile are grey-brown (palest below, with a spotted throat and barred belly) and resembles a female Blackbird. Alarm call is a sharp *tak-tak*; male sings a loud, melodious song from rocky vantage point. Favours rocky coasts and mountains. Shy. Widespread resident from Europe to Middle East, and NW Africa; found on most islands. Altitudinal and short-range migrant within range.

4. CETTI'S WARBLER *Cettia cetti* 13.5cm

Unobtrusive warbler that has rather undistinguished plumage but that is noted for its loud, explosive song. Upperparts are mostly dark brown while underparts are mainly pale greyish buff; note the pale supercilium and the rather long, broad and rounded tail (often raised). Calls include a loud *tchrr-tchrr* and a sharp *tchhtt*. Song is a loud *chee-chippi-chippi-chippi*. Usually feeds, calls and sings from within cover. Favours marshes and bushy cover close to water. Widespread and common resident but absent from NE Africa and many uplands.

5. ZITTING CISTICOLA (FAN-TAILED WARBLER)
Cisticola juncidis 10cm

Tiny, wren-sized bird with streaked brown plumage, resembling a miniature Sedge Warbler. Upperparts are sandy-buff with dark streaks on the crown, mantle and wings; underparts are pale buff, suffused buffish brown on the breast. Note the hint of a pale supercilium and the relatively long, fan-shaped tail, which is brown with a black and white margin. Call is a sharp *chit*. Song comprises an endlessly repeated *tzit-tzit-tzit…*, delivered in bouncing flight. Favours marshes, arable fields and grassland. Widespread resident in W European Mediterranean and NW Africa, becoming more local in E of region.

1. SAVI'S WARBLER *Locustella luscinioides* 14–15cm

Unobtrusive warbler. Difficult to observe because of favoured habitat but song is distinctive. Upperparts are uniformly unmarked reddish brown; underparts are unmarked pale buff, flushed rufous on breast and flanks. Long undertail coverts (typical of *Locustella* warblers), are reddish buff. Call is a sharp *svich*; song is insect-like, often heard after dark. Favours reedbeds. Passage migrant and widespread but local summer visitor (May–Aug).

2. COMMON GRASSHOPPER WARBLER *Locustella naevia* 13cm

Skulking and secretive warbler; migrants sometimes sing (from cover), making them easier to pinpoint. Upperparts are olive-brown to olive-buff with dark streaks on back and dark feather centres on wings; underparts are pale but flushed buffish brown on breast. Note the long undertail coverts, which are pale but marked with dark arrowhead streaks. Call is a sharp *tsvit*; song is mechanical and insect-like and often delivered at dusk or after dark. Migrants typically favour areas of dense grassy cover. Widespread passage migrant and local summer visitor to N.

3. RIVER WARBLER *Locustella fluviatilis* 15–16cm

Dark and plain *Locustella* warbler. Unobtrusive and usually remains well hidden. Spring migrants sometimes sing on passage, making them easier to pinpoint. Upperparts uniform dark olive-brown; throat and breast buffish brown with dark streaks, and flanks and vent have a brown wash. Note the long undertail coverts, which are dark with pale feather tips. Call is a soft *tshik*; song is an endless, mechanical buzzing *tzi-tzi-tzi…*, insect-like and recalling 'song' of field crickets (*Gryllus* sp.). Migrants typically favour dense scrub near water. Passage migrant through E of region; local summer visitor to NE of region.

4. GRACEFUL PRINIA *Prinia gracilis* 11cm

Delightfully scruffy bird. Resembles Scrub Warbler, with compact, wren-like body and relatively long, graduated tail. Note the paler unmarked underparts and absence of pale supercilium. Upperparts are greyish buff with dark streaking on back and crown; underparts are whitish. Tail is long and mostly greyish buff, but with white and dark grey patches at the feather tips. Typical call is a whistling trill; song is a series of repetitive whistles. Favours areas of scrub and tall grass, often near water. Resident from S Turkey to Middle East.

5. SCRUB WARBLER *Scotocerca inquieta* 11cm

Compact and tiny bird, recalling *Sylvia* warbler. Has wren-like body proportions but a long, fan-shaped tail that is often cocked up and splayed; tail is mostly dark but tipped white. Upperparts are greyish buff and faintly streaked; underparts are pale with faint streaking on breast and pinkish flush on flanks. Note the pale supercilium, most striking and palest in birds from the Middle East. Utters whistling and trilling calls, elements of which are included in song. Favours steppe and semi-desert habitats. Resident in N Africa and the Middle East.

6. MOUSTACHED WARBLER *Acrocephalus melanopogon* 14cm

Similar to Sedge Warbler but note the very dark cap, long pale supercilium (which ends abruptly, not tapering) and dark eyestripe grading to grey cheeks. Back is streaked and brown, and underparts are whitish with sandy-brown flanks. Calls include a harsh *trrrr*; song is scratchy but musical, often with wader-like piping notes. Favours reedbeds. Widespread but local resident; some birds (especially in E) disperse or migrate S in winter.

1. CLAMOROUS REED WARBLER *Acrocephalus stentoreus* 16–18cm

Similar to Great Reed Warbler. Upperparts are dark brown and unmarked; underparts are pale olive-brown and unmarked, palest on the throat and belly. Bill is more slender than that of Great Reed Warbler. Calls and song are loud and similar to Great Reed Warbler, with frog-like croaks. Favours tall wetland vegetation such as papyrus and reeds. Rather local resident in the Middle East.

2. GREAT REED WARBLER *Acrocephalus arundinaceus* 19–20cm

Large and robust warbler. Presence readily detected by loud and distinctive song. Upperparts are sandy-brown, darkest on the crown and most rufous on the rump; underparts are pale buff with rufous wash on the flanks. Note also the pale supercilium and relatively long, stout bill. Call is a harsh *tchak*; song is loud (with frog-like croaks and high-pitched notes), typically *krr-kree, korr-korr-korr, tsee-tsee-tsee....* Favours dense reedbeds and marshy river margins. Widespread summer visitor and passage migrant, present Apr–Sept.

3. MARSH WARBLER *Acrocephalus palustris* 12–13cm

Superficially similar to Reed Warbler. Most easily located and identified by song. Compared to Reed Warbler, has paler throat, yellowish-buff underparts and pinkish (not reddish-brown) legs. Call is similar to that of Reed Warbler; song contains plenty of mimicry of European songsters and species from African wintering grounds in addition to Reed Warbler-like elements. Favours stands of herbaceous plants (such as stinging nettles) on wetland margins. Scarce passage migrant and local summer visitor, easiest to observe in E.

4. EUROPEAN REED WARBLER *Acrocephalus scirpaceus* 12–13cm

Familiar *Acrocephalus* warbler. Upperparts are unmarked and sandy-brown, with a hint of rufous on the rump; underparts are pale (some birds have a buffish wash on the flanks). Calls include a short *tche*; song contains grating and chattering elements and some mimicry (phrases are often repeated two or three times); often delivered from a reed stem. Favours reedbeds for breeding but wetland margins generally on migration. Common and widespread passage migrant and locally common breeder, present Apr–Sept.

5. SEDGE WARBLER *Acrocephalus schoenobaenus* 13cm

Well-marked warbler. Upperparts of adult are sandy-brown and streaked; underparts are pale buff, suffused warm buff on breast. Head shows dark eyestripe and pale supercilium. Immature birds are similar to adult but breast has faint dark streaks. Call is a sharp *chek*; song is similar to that of Reed Warbler but is harsher and scratchier. Favours wetlands. Widespread passage migrant throughout the region, and local breeding species, present Apr–Sept.

6. AQUATIC WARBLER *Acrocephalus paludicola* 12–13cm

Superficially similar to Sedge Warbler but distinguishable at all times by the pale central stripe on the otherwise dark crown. Adult upperparts are streaked sandy-brown, the mantle defined by pale stripes; underparts are pale, flushed buff and streaked on the breast. Note the pale supercilium and dark eyestripe. Immature birds are similar to adult but with unmarked underparts. Call is a sharp *tchek*; song is similar to that of Sedge Warbler but subdued. Favours wetland margins. Rather scarce passage migrant through E of region.

1. OLIVACEOUS WARBLER *Hippolais pallida* 12–13cm

Tiny, delicate warbler. Recently split into two species: **Western Olivaceous Warbler** *H. opaca* (breeds S Iberia and NW Africa), which is marginally larger and darker than **Eastern Olivaceous Warbler** *H. pallida* (breeds in E of region); treated as a single species here. Adult has pale olive-brown to olive-grey upperparts and creamy-white underparts; flanks are washed buff. Head has an indistinct pale supercilium and a flat crown. Bill is slender. Immature is similar to adult but upperparts are warmer brown with a deeper buff wash to the flanks. Call is a sharp *tchak*; song is a scratchy warble, recalling those of Garden and Reed Warblers. Favours scrub areas including farmland hedgerows and gardens. Passage migrant and locally common summer visitor (Apr–Sept); may be resident in part of African range.

2. UPCHER'S WARBLER *Hippolais languida* 14–15cm

Similar to Olivaceous Warbler, but note the greyer overall appearance, the stouter bill and the hint of a pale panel on the wings. Adult upperparts are grey to grey-buff and underparts are whitish with a very pale buffish wash on the flanks. Note also the pale supercilium and rather dark tail. Immature birds are similar to adult but colour is overall warmer brown. Call is a sharp *tsak*; song is a scratchy warble. Sometimes swings fanned tail from side to side. Favours areas of scrub with taller trees. Summer visitor and passage migrant to the Middle East and E Turkey, present Apr–Sept.

3. MELODIOUS WARBLER *Hippolais polyglotta* 13cm

Rather colourful warbler, recalling an outsized Willow Warbler but with a proportionately larger bill. In spring and summer, adult upperparts are greenish brown and underparts are yellow. Note also the pale yellow lores. In autumn, adult and immature birds are typically duller in colour. Calls include a harsh *tchret*; song is rich and warbling. Favours wooded scrub areas, often close to water. Passage migrant and summer visitor to W of region, present Apr–Sept.

4. OLIVE-TREE WARBLER *Hippolais olivetorum* 16–17cm

A large and robust warbler, similar in size to Great Reed Warbler. In spring and summer, adult has grey-brown upperparts and dirty white underparts, with a subtle darker wash to the breast. Bill is relatively long and stout, and note also the pale supercilium in front of the eye, the pale panel on the wings and the dark legs. In autumn, adult and immature birds have a warmer buff tone to overall plumage colours than spring adult. Call is a sharp *tchek*; song is warbling with harsh, croaking elements, recalling that of Great Reed Warbler. Favours mature olive groves and areas of maquis with mature trees. Summer visitor to E of region, mainly Greece and Turkey, and passage migrant through the Middle East, present Apr–Sept.

5. ICTERINE WARBLER *Hippolais icterina* 12–13cm

Similar to Melodious Warbler but distinguishable with care. Note the hint of a pale wing panel, the longer primary projection (in folded wings) and the grey (not reddish-brown) legs. In spring and summer, adult has greenish-brown upperparts and yellow underparts. Note the pale yellow lores and hint of a pale supercilium. In autumn, adult and immature birds are duller than spring adult. Calls include a loud *tee-too-lueet*; song is rich and full of mimicry. Favours scrubby woodland and even occurs in overgrown gardens. Passage migrant, commonest in E of region, and local breeding species from E France to Bulgaria northwards, present Apr–Sept. Little overlap, in terms of breeding range, with Melodious Warbler.

1. DARTFORD WARBLER *Sylvia undata* 12–13cm

Compact warbler with needle-like bill and long tail that is often cocked up. Adult male has blue-grey upperparts, reddish-maroon breast and flanks, and white down centre of belly. Note red eye-ring surrounding dull red eye, yellow base to bill, and reddish-yellow legs. Adult female is similar to male but reddish elements of plumage are duller. Immature is similar to adult female but mantle and wings are brownish and underparts are dull grey. Call is a harsh *tchrr* or *tchrrr-tt*; song is fast and scratchy. Favours maquis and heaths. Resident Iberia and S France to Italy (also local NW Africa). Some dispersal in winter.

2. MARMORA'S WARBLER *Sylvia (sarda) sarda* **and** BALEARIC WARBLER *S. (s.) balearica* 11–12cm

Formerly a single species. Adult male Marmora's has dull blue-grey plumage, darkest on wings and tail and palest on the underparts; note the reddish legs, the red eye-ring, the dark 'mask' and the orange-yellow base to the bill. Adult male Balearic is similar but with paler underparts, particularly the throat. Adult females and immatures (of both species) are similar to their respective males but duller overall and browner, and the throat paler. Call is a harsh *chak* (Marmora's) and *tcheck* (Balearic); scratchy song is similar to that of Dartford Warbler. Favours garrigue and low maquis. Balearic is resident on Mallorca, Ibiza and Formentera, Marmora's mainly on Corsica and Sardinia. Some winter dispersal, when range extends to NW Africa.

3. DESERT WARBLER *Sylvia nana* 12cm

Adult from NW Africa (*S. n. deserti*) has pale sandy-brown upperparts and whitish underparts; note the pale yellow iris, the dark-tipped yellow bill and the yellow legs. Adult from E of region (*S. n. nana*) is similar to W adult but upperparts are pale grey-brown. Chattering call is similar to that of a Blue Tit; song is fast and musical. Favours steppe vegetation and semi-desert. Resident in NW Africa; winter visitor and passage migrant to Middle East (Nov–Mar).

4. TRISTRAM'S WARBLER *Sylvia deserticola* 12cm

Adult male has blue-grey upperparts and reddish-orange underparts; note white eye-ring, reddish eye, chestnut margins to wing feathers and hint of a pale moustachial stripe. Adult female resembles male but underparts (including the throat) are pale orange-buff and the whitish moustachial stripe is prominent. Immature has grey-buff upperparts, chestnut margins to wing feathers, a white throat and otherwise pale pinkish-buff underparts. Call is a sharp *tchett*; song is fast and scratchy. Favours scrubby areas. Resident.

5. BARRED WARBLER *Sylvia nisoria* 15cm

Adult male has blue-grey upperparts, dark wings with two pale wingbars and dark-barred whitish underparts, Note the yellow eye and white-tipped tail. Adult female is similar but colours and barring are more subdued. Immature is grey-brown above, pale buff below, with two pale wingbars and faint barring on tail coverts. Calls include a sharp *tak*. Favours damp scrub. Passage migrant and local summer visitor to E (Apr–Aug).

6. LESSER WHITETHROAT *Sylvia curruca* 13–14cm

Has blue-grey crown, dark mask, and brownish-grey back and wings; throat and underparts white with faint buffish wash on flanks. Male brighter than female or immature, sometimes with pink flush to the breast. Call is a sharp *chek*; rattling song is preceded by soft warbling. Widespread passage migrant and summer visitor, especially to E of region, present Apr–Sept.

1

2

2 ♂

Balearic

3. *nana*

2. Balearic

4 ♂

deserti 3

4 ♂

4

5. imm

5

6

1. GARDEN WARBLER *Sylvia borin* 14cm

A rather nondescript warbler with few distinctive characteristics. Sexes are similar. Both adult and juvenile have uniform grey-brown upperparts and paler, buffish underparts. Note the dark eye, the faint pale supercilium, the rather short bill and the indistinct grey feathering on the side of the neck. Lack of distinguishing plumage features is made up for by the male's attractive and musical song, which is similar to that of Blackcap; call is a harsh *tchek-tchek*. Favours wooded areas with dense undergrowth. Widespread passage migrant throughout, and local summer visitor, mainly to European parts of Mediterranean, present Apr–Sept.

2. BLACKCAP *Sylvia atricapilla* 14cm

Distinctive warbler. Adult male has grey-brown upperparts, paler greyish underparts (palest on the throat and undertail) and a distinctive black cap. Note also the indistinct pale eye-ring. Female and juvenile are similar to male but have a chestnut-brown (not black) cap. Male's song is attractive and musical with chattering and fluty elements; call is a sharp *tekk-tekk*. Occurrence and distribution is complex. Widespread passage migrant throughout. However, residents and winter visitors ensure year-round occurrence in much of Europe and NW Africa, although in NE, typically a local summer visitor.

3. SPECTACLED WARBLER *Sylvia conspicillata* 12.5cm

Resembles a small Whitethroat. Adult male has a dark grey head, a sandy-grey back and tail, and chestnut fringes to the wing feathers; throat is strikingly white but upperparts are otherwise pinkish grey. Note also the white eye-ring. Female resembles the male but has paler plumage colours overall, a brownish (not grey) head and an inconspicuous eye-ring. Juvenile is similar to female but upperparts are warmer sandy-brown. Call is a rattling *drrr*; song is fast and warbling. Favours low maquis, garrigue and areas of glasswort. Local summer visitor to W European Mediterranean, present Apr–Sept; resident and winter visitor to N Africa and the Middle East.

4. SUBALPINE WARBLER *Sylvia cantillans* 12cm

One of the most typical Mediterranean warblers. Adult male has blue-grey upperparts except for the brownish wings. Throat, breast and flanks are orange-red, while the belly and undertail are white. Note the red eye-ring and the striking white moustachial stripe. Adult female and juvenile resemble adult male but plumage colours are subdued, with particularly pale underparts; a hint of a pale moustache is usually still discernible. Call is a rasping *chett*; song is a fast and chattering musical warble. Favours maquis and woodland clearings. Widespread passage migrant and summer visitor from Iberia to W Turkey, and NW Africa, present Apr–Sept.

5. COMMON WHITETHROAT *Sylvia communis* 14cm

Widespread and familiar warbler. Adult male has blue-grey crown, a grey-brown back and rufous wings; throat is strikingly white but underparts are otherwise greyish, suffused with a buffish tone on the breast. Note the white eye-ring. Female and juvenile are similar to male but with subdued plumage colours and brownish (not grey) crown. Call a harsh *chek*; song is a scratchy warble, often delivered in a dancing song flight. Favours scrub, heaths, maquis and farmland with hedgerows. Widespread passage migrant and summer visitor, present Apr–Sept.

1. MÉNÉTRIES' WARBLER *Sylvia mystacea* 12–13cm

Adult male has dark grey cap and otherwise grey upperparts; underparts are whitish (palest on throat) but E birds are flushed pinkish on throat and breast; note red eye and yellow eye-ring. Female has sandygrey upperparts and pale underparts, palest on throat and washed buff on breast and flanks. Call is a sharp *tchak*; song is fast and warbling. Favours scrub and maquis. Passage migrant in Middle East; summer visitor E from SE Turkey (Apr–Sept).

2. SARDINIAN WARBLER *Sylvia melanocephala* 13.5cm

Adult male has black cap, striking white throat and red eye and eye-ring; upperparts are otherwise grey and underparts are greyish white. Female and juvenile recall male but cap is greyish, eye colours are subdued, and body is brownish, darkest on upperparts. All birds have reddish-yellow legs. Calls include a rattling *chett-chett-chett-chett*; song is a fast warble. Favours scrub and open woodland. Resident or partial migrant across much of range, mainly migratory in E. Widespread in N Africa and the Middle East in winter.

3. CYPRUS WARBLER *Sylvia melanothorax* 13.5cm

Adult male has black cap, white moustache and grey upperparts; underparts are pale grey with black spots; note reddish iris and pale eye-ring. Adult female resembles male but is duller and browner overall, with a grey cap; spots on underparts are paler and restricted to throat and breast. Juvenile has grey cap and otherwise grey-brown upperparts; underparts are unspotted buffish grey, palest on throat. Call is a sharp *chek*; song is a coarse warble. Favours maquis and scrub. Breeds only on Cyprus (Apr–Aug). Passage migrant and winter visitor to Middle East.

4. RÜPPELL'S WARBLER *Sylvia rueppelli* 14cm

Adult male has black head with white moustache and bright red eye and eye-ring; upperparts are otherwise grey and underparts are greyish white. Female recalls male but plumage is typically paler; throat either pale or dark-spotted. Immature is similar to female but throat is pale. All birds have reddish legs. Call is a rattling *chrrr*; song is a fast warble. Favours maquis. Summer visitor, mainly Greece and Turkey (Apr–Aug); passage migrant in Middle East.

5. ORPHEAN WARBLER *Sylvia hortensis* 15cm

Recently split into: **Western Orphean Warbler** *S. (h.) hortensis* and **Eastern Orphean Warbler** *S. (h.) crassirostris*. Western adult male has dark hood grading to grey-brown upperparts. Note whitish iris, white throat, and whitish underparts flushed pinkish buff on breast and flanks; undertail coverts are uniform buff. Eastern adult male resembles Western male but upperparts are greyer and underparts paler; undertail coverts are pale but dark-tipped. Female and juvenile (of both species) are similar to respective males but with browner plumage overall and brownish-grey hood; iris is dull, darkest in juvenile/first-autumn birds. Call is a harsh *tekk*; song is warbling and rich. Favours open woodland, olive groves and scrub. Passage migrant and widespread summer visitor, present Apr–Sept.

6. ARABIAN WARBLER *Sylvia leucomelaena* Length 15–16cm

Recalls Orphean Warbler but note dark eye with whitish eye-ring. Black tail is rather long, white-tipped below and habitually flicked down. Male has unmarked white underparts, grey back and blackish cap. Female and immature recall male but upperparts are browner. Song is a rich warble and call is a harsh *tchak-chak*. Only in ancient acacias. Resident in S Israel.

1

2 ♀

2 ♂

2 ♂

3

4 ♀

4 ♂

5. 1st-w

6 ♂

5 ♂

6 ♂

1. CHIFFCHAFF *Phylloscopus collybita* 11cm

Typical adult has dull olive-brown upperparts, pale yellow under-parts and buff wash on flanks. E birds are typically greyer than W birds. Juveniles are usually yellowish. All birds have black legs. Song is a variation on the *chiff-chaff* theme; calls include a soft *hu-eet*. Favours woodland and scrub. Common passage migrant and summer visitor. Occurs year round in much of Europe; numbers boosted and range extends S and E in winter.

2. IBERIAN CHIFFCHAFF *Phylloscopus ibericus (brehmii)* 11cm

Formerly treated as a race of Chiffchaff. Compared to that species, adult is warmer yellow-buff, particularly on the underparts, wings are slightly longer and legs are slightly paler. Juvenile is often suffused with yellow. Call is a soft *hu-eet*; song contains both Chiffchaff and Willow Warbler elements. Favours woodland and scrub. Resident in N Iberia, S France and NW Africa.

3. WESTERN BONELLI'S WARBLER *Phylloscopus bonelli* and EASTERN BONELLI'S WARBLER *P. orientalis* 11.5cm

Formerly a single species. Western Bonelli's has pale grey-brown upperparts and whitish underparts, including the throat; note the greenish-yellow rump, wing patch and margin to outer tail feathers. Eastern Bonelli's is similar but plumage is duller and greyer above and greenish-yellow elements of plumage are less striking. Both species deliver a trilling song: call of Eastern Bonelli's is a sparrow-like *chirp*, while that of Western Bonelli's is a softer *hu-eet*. Both species favour wooded slopes, often breeding in hills. Widespread passage migrants and summer visitors to their respective halves of the region, present Apr–Aug.

4. WOOD WARBLER *Phylloscopus sibilatrix* 12cm

Recognised by its olive-green upperparts, its yellow face, supercilium and throat, and its white underparts. Legs are flesh-coloured in all birds. Call is a sharp *tsip*; song is an accelerating, trilling warble. Favours open woodland, especially beech. Widespread but generally scarce passage migrant and local breeding species, present Apr–Aug.

5. WILLOW WARBLER *Phylloscopus trochilus* 11cm

Adult upperparts are olive-yellow and underparts are pale yellowish white; juvenile is similar but brighter and more yellow. Legs are flesh-coloured in all birds. Song has tinkling, descending phrases; call is a soft *hoo-eet*. Widespread passage migrant and local summer visitor, mainly Apr–Aug.

6. GOLDCREST *Regulus regulus* 9cm

Tiny bird with a needle-like bill and large eye and head. Upperparts are greenish with two pale wingbars; underparts are grey-buff. Black-bordered crown is orange in male, yellow in female; juvenile lacks adult's crown markings. Call is thin and high-pitched; song comprises five or six thin notes followed by a high-pitched trill. Favours conifer woodland. Local resident and more widespread winter visitor, mainly N Iberia to Turkey northwards.

7. FIRECREST *Regulus ignicapilla* 9cm

Adult has olive-green upperparts with a bronze patch on side of neck, two pale wingbars and a white supercilium; underparts are whitish. Male has black-bordered orange-yellow crown stripe; crown stripe of female is yellow. Call is similar to that of Goldcrest; song is a series of thin notes. Favours mature mixed and conifer woodland. Locally common resident and winter visitor.

1. SEMI-COLLARED FLYCATCHER *Ficedula semitorquata* 13cm

Similar to both Pied and Collared flycatchers. With adult male in spring and early summer, note the small white forecrown patch (like Pied), the white half-collar (more extensive than on Pied) and the large amount of white on the wings (like Collared, but with white on median coverts as well). Rump is pale, midway between Pied and Collared in terms of extent of white. Adult female is similar to female Collared but the white-tipped median coverts form a second wingbar. By late summer, adult male resembles female (but with more white on wings); first-autumn bird is similar to adult female. Call is a whistling *peeup*; song comprises a series of whistling phrases. Scarce passage migrant in E of region and local breeding species eastwards from Greece, present Apr–Aug.

2. PIED FLYCATCHER *Ficedula hypoleuca* 13cm

Distinctive flycatcher. In spring and early summer, adult male has mainly black upperparts and white underparts. Note the small white forecrown patch, the large white wing patch and the small amount of white at the base of the primaries. Female has a similar pattern to the male but black elements of plumage are replaced by brown, and white on the wings is less extensive. By late summer (after moulting) adult male resembles female (but with more white on the wings); first-autumn bird is very similar to adult female. Call is a sharp *tek*; tuneful song comprises several phrases, each of which is typically repeated two or three times. Catches insects in flight, launching forays from repeatedly used perch. Favours deciduous woodland, especially oak, but migrants often linger in wooded parks and olive groves. Widespread passage migrant (easiest to see in spring), and local breeding species in W Europe and NW Africa, present Apr–Sept.

3. COLLARED FLYCATCHER *Ficedula albicollis* 13cm

Similar to Pied Flycatcher. Adult male in spring and early summer has mainly black upperparts and white underparts. Note the larger white forecrown patch than on Pied, the striking white collar, and the white rump and more extensive white on the wings, especially at the primary bases. Female is similar to female Pied, but note the greyer upperparts and greater extent of white at the primary bases. Adult male in late summer resembles female (but with more white in the wings); first-autumn birds resemble female closely. Call is a distinctive *eerp*; song comprises a series of thin whistling phrases. Passage migrant through E of region and local breeding species from Italy and Greece northwards, present Apr–Aug.

4. RED-BREASTED FLYCATCHER *Ficedula parva* 11.5cm

Attractive flycatcher. Adult male has greyish head and brown upperparts; throat and upper breast are reddish but underparts are otherwise white. Note also the white sides to the black tail. Adult female resembles male but has pale throat and brownish head. First-autumn bird is similar to female but throat is buffish. Frequently flicks tail. Calls include a wren-like rattle; song comprises thin, musical phrases. Favours wooded areas. Scarce passage migrant, and local summer visitor, mainly in E of region, present mainly Apr–Aug.

5. SPOTTED FLYCATCHER *Muscicapa striata* 14cm

Adult has grey-brown upperparts, streaked on crown, and pale underparts, heavily streaked on the breast. Juvenile is similar but with a dark-spotted breast and buff-spotted back. Adopts upright stance and makes insect-catching sorties from a perch. Calls include a soft *tseer*; song comprises a few call-like phrases. Common passage migrant and summer visitor (Apr–Sept).

1

2

2

2

3

4

4. 1st-autumn

1st-autumn

5

5

1. CRESTED TIT *Parus cristatus* 11–12cm

Distinctive bird with striking black and white crest. The white face is marked by a dark stripe through the eye and a dark border around the ear coverts. Shows a black throat and collar, but upperparts are otherwise brown and underparts are pale. Note also the beady red eye. Presence is often indicated by a trilling call; song comprises a series of call-like notes. Favours pine, Cork Oak and Beech woodland. Resident in Iberia, S France and Greece.

2. GREAT TIT *Parus major* 14cm

Striking tit. Adult has a black head with a large white cheek patch, and black bib that forms a line (broader in male than female) running down the chest. Underparts are otherwise yellow and upperparts are mainly greenish; note the white wingbar. Juvenile plumage pattern resembles that of adult but colours are washed out and there is no white on the head. Calls include a ringing *tchink-tchink*; typical song is a loud *teecha-teecha-teecha*. Widespread woodland resident although absent from NE Africa and local in Middle East.

3. BLUE TIT *Parus caeruleus* 11–12cm

Colourful songbird. Typical adult has blue wings, a green back and yellow underparts. Head is mainly white but marked with striking dark blue lines and a blue cap. Adults from N Africa are similar but back is blue, while cap and facial lines are very dark blue (almost black). Juvenile is similar to adult but blue elements of adult's plumage are greenish. Call is a chattering *tserr-err-err-err*; song is whistling. Favours broadleaved (generally deciduous) woodland. Widespread resident; absent from NE Africa and Middle East.

4. COAL TIT *Parus ater* 11–12cm

Warbler-like tit. Typical adult has mainly black head with striking white cheek patch and narrow white patch on nape. Back and wings are slate-grey with two white wingbars; underparts are pinkish buff. Cyprus birds are darker overall with a brown back, rufous-buff underparts and a more extensive black bib. Birds from NW Africa have the back, underparts and cheek washed with yellow. Calls include a thin *tseep*; song is a high-pitched *tseechu-tseechu-tseechu….* Favours conifer forests. Widespread resident, but absent NE Africa and Middle East.

5. MARSH TIT *Parus palustris* 12cm

Plump-bodied tit with a relatively large head and a stubby little bill. Smaller than Sombre Tit, with warmer brown plumage and a smaller bib. Has a black cap (appears glossy in good light), white cheeks and a small black bib. Upperparts are otherwise brown and underparts are buffish, darkest on the flanks. Call is a loud *pitchoo*; song is a repetitive series of notes such as *chip-chip-chip….* Favours broadleaved woodland. Resident from N Spain to W Turkey. **5a. Willow Tit** *P. montanus* is very similar to Marsh Tit, but note dull black cap, larger black bib and pale panel on secondaries. Calls include a harsh *si-si-tcha-tcha-tcha*. Restricted in the region to upland woodlands.

6. SOMBRE TIT *Parus lugubris* 13–14cm

Comparatively large tit. Striking head pattern comprises a dark brown cap and an extensive dark brown bib, separated by a broad whitish cheek patch. Upperparts are otherwise grey-brown and underparts are whitish, washed with greyish buff. Calls include a Blue Tit-like *tsi-tsi-tchrrr*; song comprises a simple series of whistling notes. Favours open woodland with stone walls, or maquis with rocky outcrops. Resident in E, mainly Greece and Turkey.

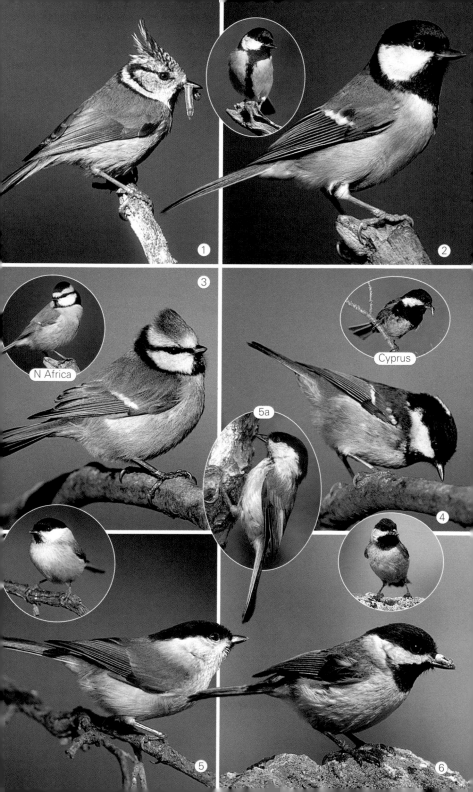

N Africa

Cyprus

5a

1

2

3

4

5

6

1. LONG-TAILED TIT *Aegithalos caudatus* 14cm

Dumpy bird with a stubby bill and a very long tail. Across most of range, adult has a whitish head with a broad, dark stripe along the side of the crown. Underparts are whitish with a pinkish wash to the flanks and belly. Upperparts are mainly black but note the reddish-chestnut patches on the sides of the mantle and the white wing feather margins. Adults from S Iberia are similar but mantle colour is grey (not chestnut) and cheeks and throat have dark streaking. Adults from E of region have grey on the mantle, a streaked face, a black bib and greyish-white underparts. Juvenile is mostly whitish below and grey-brown above; crown is white but sides of head are brown. Calls include a wren-like trill; song is trilling and subdued. Often seen in flocks. Favours woods and scrub. Resident in Europe and Turkey.

2. BEARDED REEDLING *Panurus biarmicus* 14–16cm

Attractive and distinctive wetland bird with a proportionately long tail. Adult male has a blue-grey head and a black 'moustache'; note also the yellow bill and yellow iris. Upperparts are otherwise orange-buff (with some dark-centred wing feathers and whitish flight feathers) while underparts are pale buff. Adult female resembles male, but head is buffish, grading to pale grey on the throat, and it lacks the male's 'moustache'. Juvenile female is similar to adult female but eye colour is dull. Juvenile male resembles adult female but has a dark patch on the back and a hint of a dark mask. Call is a distinctive *ping*; squeaky song is seldom heard. Forms flocks outside the breeding season. Favours extensive reedbeds. Local resident in Europe and Turkey; some dispersal occurs in winter.

3. PENDULINE TIT *Remiz pendulinus* 11cm

Tiny and compact-bodied bird with a narrow and fine, triangular bill. Adult plumage is reminiscent of that of a male Red-backed Shrike. Has a grey head, a black mask and a chestnut back; the flight feathers and tail are black and the underparts are whitish with a buffish wash on the flanks. Sexes are similar but male has a broader black mask than female and is flushed with reddish brown on the breast. Juvenile has subdued plumage colours (brownish above and pale buff below) and no face mask. Call is a thin, whistling *tseee*; trilling song includes call-like phrases. Favours extensive wetlands, such as reedbeds and deltas. Unique flagon-shaped nest, made from spider's silk and seed-hairs, is suspended from a drooping branch. Local resident in Europe and Turkey; dispersal and arriving migrants mean winter range extends S and E.

4. SHORT-TOED TREECREEPER *Certhia brachydactyla* 12–13cm

Short hind claw is not a useful field indicator; easiest to identify by voice. Has streaked brown upperparts. Compared to Treecreeper, bill is longer and more downcurved and underparts show a more intense buffish wash; supercilium is pale behind the eye but grubby in front. Call is a sharp *tzeeut*; song is similar to that of Treecreeper but more abupt. Favours deciduous and coniferous woodland. Resident from Iberia to Turkey, and NW Africa.

5. EURASIAN TREECREEPER *Certhia familiaris* 13–14cm

Unobtrusive bird with a needle-like, slightly downcurved bill. Creeps up tree trunks and can look rather mouse-like. Upperparts are streaked brown and underparts are white; note that some birds show a faint brown wash to the vent. Typically has a pale supercilium in front of the eye. Calls include a high-pitched, thin *sreeit*; song is a short series of thin notes ending in a trill. Favours conifers and mixed woodland. Resident from N Spain to N Turkey northwards.

① ② ③ ④ ⑤

1. KRÜPER'S NUTHATCH *Sitta krueperi* 12cm

Similar proportions to Eurasian Nuthatch but appreciably smaller, with a proportionately smaller bill and with distinctive markings. Upperparts are blue-grey while underparts are pale greyish white in female and bluish grey in male. Note also the black eyestripe, the white supercilium and the reddish-chestnut patch on the breast; the forecrown is black, the extent of which is greater in the male than the female. Calls include a nasal *kluee*, song is a nasal *tui-tui-tui...*, reminiscent of a toy trumpet. Favours conifer forests, typically pine. Resident in Turkey and the Greek Islands of Lesvos and Khios.

2. EURASIAN NUTHATCH *Sitta europaea* 14cm

The most widespread nuthatch in the region. Like all nuthatches, it is characterised by its plump body and chisel-like bill. Upperparts are blue-grey and underparts are variably pale orange-buff; note also the black eyestripe and the white cheeks. Sexes are similar but underparts of male are typically more intense reddish buff (especially towards the vent) than on female. Calls include a loud *whett*, often repeated several times; song is a falcon-like *piu-piu-piu....* Often descends tree trunk head downwards, with typically jerky movements. Favours broadleaved and mixed woodlands. Resident across most of European Mediterranean range; also occurs in Turkey and, locally, in NW Africa.

3. CORSICAN NUTHATCH *Sitta whiteheadi* 11–12cm

Similar size and proportions to Krüper's Nuthatch but plumage pattern is distinctly different; also widely separated geographically so that no overlap occurs. Upperparts are blue-grey and underparts are pale buffish grey. Note the black crown, broad white supercilium and black eyestripe in the male; in the female, the black elements of the head pattern are blue-grey (the same colour as the back). Calls include a nasal *kshreeh*; song is a rapid *tu-tu-tu....* Favours ancient stands of Corsican Pine on forested mountain slopes. Resident on Corsica. Endemic to that island and the only nuthatch species to occur there. **Algerian Nuthatch** *Sitta ledanti* (not illustrated) is similar to Corsican Nuthatch but male has a black forecrown (not the entire crown) and underparts are cleaner looking, with a warm pinkish-buff wash; vent is white (not grey-buff) in both sexes. Confined to small areas of mountain forest in Algeria. No other nuthatch species occurs in the area and so confusion is impossible. Range is so restricted that you are most unlikely to encounter this species elsewhere by chance.

4. WESTERN ROCK NUTHATCH *Sitta neumayer* 15cm

Larger and paler than Eurasian Nuthatch with a proportionately much longer bill. Note also this species' different habitat preferences. Upperparts are pale grey and underparts are white with a variable buff wash, most intense on the belly. Note also the long and narrow black eyestripe; this narrows behind the eye, but when the head is twisted this feature is not always obvious. Calls include a shrill *kiu-kiu-kiu...*; song comprises a series of whistling notes. Favours warm, rocky terrain. Builds a large and characteristic mud nest, plastered into a rock crevice and with an entrance tunnel; despite its size, the nest is surprisingly well camouflaged. Widespread resident from eastern Adriatic, through Greece and Turkey and S to parts of Middle.East. **4a. Eastern Rock Nuthatch** *Sitta tephronota* is superficially similar to Western Rock Nuthatch but larger (length 16cm). Note also the warmer buff underparts, the even larger bill and the thicker black eyestripe, which broadens behind the eye. Favours rocky mountain slopes with scattered trees. Resident from E Turkey eastwards.

1. WOODCHAT SHRIKE *Lanius senator* 18cm

Adult male has a chestnut crown and nape; upperparts are otherwise black except for white markings on the wings and shoulder, and a white rump; underparts are white. Adult female is similar to male but colours are duller. Juvenile has grey-brown upperparts and whitish underparts, all with extensive scaling. Call is a grating *tschrrr*; song is a harsh, musical warble. Favours open country and farmland with wires and scattered scrub. Widespread passage migrant and summer visitor, present Apr–Aug.

2. RED-BACKED SHRIKE *Lanius collurio* 17cm

Male has a reddish back, pale underparts flushed with pink, a blue-grey cap and a dark mask; tail is black with white on sides at base. Female is similar but colours are less distinct; shows crescent markings on underparts. Juvenile is brown and adorned with crescent markings. Call is a sharp *chrr*; warbling song includes mimicry. Favours open country and farmland with hedgerows and scrub. Summer visitor from N Spain and France to Turkey, and passage migrant elsewhere, particularly common in E; present Apr–Sept.

3. MASKED SHRIKE *Lanius nubicus* 18cm

Elegant, slim-bodied shrike. Adult male has mainly black upperparts with white on the wings and shoulder, and a white forehead and supercilium; underparts are white with a pinkish-orange wash to flanks. Adult female is similar but colours and pattern are less distinct. Juvenile has grey-brown upperparts and whitish underparts, all with extensive scaling. Call is a harsh *chrr*; song is a musical warble. Favours olive groves, orchards and open pine woodland. Locally common summer visitor to N Greece, Turkey and Middle East, present Apr–Aug.

4. SOUTHERN GREY SHRIKE *Lanius meridionalis* 24cm

Variable across range. All birds have black mask that does not extend onto forehead. NW Africa adult has grey cap and back, black wings, and whitish-grey underparts. Note white primary bases and white on scapulars. NE Africa and Middle East adult paler overall with white supercilium. Iberia and SW France adult resembles adult from NW Africa but underparts flushed apricot; has pale supercilium. Juveniles are paler versions of respective adults; black elements of plumage are greyer. Calls include a harsh *vrrrk*; song is subdued and squeaky. Favours open country with bushes. Mainly resident (*see* resident range on map). **Great Grey Shrike** *L. excubitor* (not illustrated) is similar, with white underparts and white secondary and primary bases. Scarce winter visitor (Nov–Mar) (mainly winter range on map).

5. ISABELLINE SHRIKE *Lanius isabellinus* 17cm

Formerly treated as a race of Red-backed Shrike. Male has pale grey-brown upperparts, whitish-buff underparts, a black mask and a reddish tail; female is similar but plumage colours, other than tail, are more subdued. Juvenile is similar to juvenile Red-backed but is greyer above and has a red tail. Favours open country with bushes. Scarce passage migrant in E of region.

6. LESSER GREY SHRIKE *Lanius minor* 20cm

Male has grey crown and back, broad black eyestripe that continues across forecrown, and black wings with a broad white primary patch; throat is white but underparts are otherwise flushed pink. Female is similar to male but forecrown is often speckled. Juvenile resembles washed-out female but black forecrown is absent and back appears scaly. Favours open country. Passage migrant and local summer visitor, most numerous in E; present Apr–Sept.

1

1. juv

3

♀

2

juv

4

♀

4

5

♂

6

1. BLACK-CROWNED TCHAGRA *Tchagra senegala* 22–23cm

A striking relative of the shrikes with a powerful bill that is hooked at the tip. Head pattern is distinctive and comprises a black crown stripe, a broad creamy-white supercilium and a broad black eyestripe; face is otherwise whitish grey. Underparts are grey, back is grey-brown and wings are mainly chestnut. Tail is relatively long, dark grey, wedge-shaped and tapering; note the white spots at the tip. Song is loud and whistling; otherwise mostly silent. Favours scrub patches and generally unobtrusive. Resident in NW Africa.

2. COMMON BULBUL *Pycnonotus barbatus* 19–20cm

A noisy, superficially thrush-like bird with a relatively long, dark tail. Upperparts are a rich greyish brown, darkest on the face and throat. Breast and belly are pale grey-brown and the vent is white. Calls include a rasping *tcharr*; song is loud and bubbling. Favours orchards, gardens and oases, usually where water is present. Often seen in small flocks. Resident in NW Africa and the Nile Valley.

3. SPECTACLED (YELLOW-VENTED) BULBUL *Pycnonotus xanthopygos* 19–20cm

A distinctive, thrush-like bird. Upperparts are mainly grey-brown except for the head and tail, which are dark chocolate-brown; note the striking white eye-ring. Breast and belly are pale grey-buff while vent is bright yellow. Call is a rasping *tcharr*; loud and bubbling song is similar to that of Common Bulbul. Favours gardens, orchards, palm plantations and wadis. Often seen in small flocks. Resident from SE Turkey to the Middle East.

4. PALESTINE SUNBIRD *Nectarina osea* 11cm

A tiny, extremely active bird with a relatively short tail and a characteristic downcurved bill. Difficult to confuse with any other species in the region on account of its structure and habits. Male in breeding plumage (Dec–June) often looks jet-black; in good light, however, note the glossy blue and purple sheen. Adult female (at all times) has grey-brown upperparts and grey underparts. Non-breeding male (July–Nov) is similar to female but with scattered darker feathers. Calls include loud whistles and sharp clicks; song comprises a series of whistles, ending in a trill. A nectar-feeder, visiting flowers in gardens, parks and orchards. Local resident in the Middle East.

5. FULVOUS BABBLER *Turdoides fulvus* 24–25cm

A thrush-sized bird with a relatively long tail and a pointed and slightly downcurved bill. Upperparts are warm sandy-buff; throat is whitish but underparts are otherwise sandy-brown, the colour most intense on the flanks. Calls include mechanical trills; song comprises a series of thin whistles. Favours arid terrain and semi-deserts with scattered bushes. Usually seen in small flocks. Resident in NW Africa.

6. ARABIAN BABBLER *Turdoides squamiceps* 25–27cm

A thrush-sized bird with a relatively long tail and a downcurved bill. Upperparts are sandy-buff with a grey wash to the head; underparts are whitish but the breast and flanks are washed with greyish brown and marked with faint spots and streaks. Iris is pale in adult male but darker in female and juvenile. Calls include a sharp trill; song comprises a series of whistles. Usually seen in flocks. Favours semi-desert terrain with scattered bushes. Resident from S Israel southwards through Arabian Peninsula.

1. TRISTRAM'S STARLING *Onychognathus tristramii* 25–26cm

A Common Starling-sized bird but with a proportionately longer tail. Male has mainly glossy black plumage; note the chestnut-orange primary patch, which is most striking in flight. Female and juvenile birds have male's chestnut primary patch but plumage is otherwise blackish brown, grading to grey on the head and neck. Calls and song include shrill and grating whistles. Favours wadis and semi-desert terrain with scattered scrub. Usually found in flocks. Resident from S Israel southwards to Arabian Peninsula.

2. SPOTLESS STARLING *Sturnus unicolor* 22cm

Superficially very similar to Common Starling but distinguishable with care. Adult has mainly uniformly dark plumage at all times; in summer has a purple sheen and relatively long, shaggy throat feathers, while in winter note the subdued pale tips to the feathers on the head, back and underparts. Bill is yellow in summer but black in winter. Juvenile is similar to juvenile Common Starling but darker. Calls and song are similar to those of Common Starling. Resident in Iberia, SW France (very local), Corsica, Sardinia, Sicily and NW Africa. Usually the only starling species present within much of its range during the breeding season although Common Starling's winter range extends to these areas.

3. COMMON STARLING *Sturnus vulgaris* 22cm

Generally the most common and widespread starling species in the region. Adult has mainly dark plumage; in summer this is essentially unspotted and has a green sheen, while in winter the feathers acquire numerous white or buff spots (note also the brownish margins to the wing feathers at this time). Bill is yellow in summer but dark in winter. Juvenile is buffish brown; acquires adult-like plumage (dark with pale spots) in first autumn. Calls include a rasping *tchair*; varied song includes clicks and whistles; also imitates other birds and man-made sounds. Forms large flocks outside the breeding season. Widespread resident from S France to Turkey northwards; numbers increase in winter months and range then extends to Iberia, N Africa and the Middle East.

4. ROSE-COLOURED STARLING *Sturnus roseus* 20–22cm

Attractive and distinctive starling. Adult male in summer has a dark head and blackish wings and tail; back and underparts are otherwise pink. Note also the shaggy crest and the pink bill and legs; in good light, head has a purplish sheen and wings have a greenish sheen. Adult female in summer is similar to male but pink elements of plumage are duller. Adults in winter resemble summer adults but pink elements of plumage are grubby and the dark feathers have pale fringes. Juvenile is pale sandy-brown; note the yellow bill (dark in both Common and Spotless Starling juveniles). Passage migrant and erratic summer visitor, mainly to the E of the region, present May–Oct.

5. EURASIAN GOLDEN ORIOLE *Oriolus oriolus* 24cm

Unmistakable thrush-sized bird. Male has mainly bright yellow plumage with black wings, black on the tail and a small black 'mask'; note also the red bill. Female has a similar plumage pattern to the male but yellow elements of plumage are duller yellow-green, palest on the underparts and also streaked there. Juvenile resembles female but colours and contrast are even more subdued and bill colour is dull pink. Well camouflaged in dappled foliage. Calls include a Eurasian Jay-like scream; song is a fluty, tropical-sounding *wee-lo-weeoo*. Favours woods and copses. Widespread summer visitor and passage migrant, present Apr–Sept.

1st-year ♀

3. winter

3. juv

4. juv

1

2

3

4

5

1. AZURE-WINGED MAGPIE *Cyanopica cyana* 35cm

Colourful and distinctive long-tailed bird. Adult has a black cap, a pinkish-brown back and rump, and buffish-white underparts; wings and tail are azure-blue. Juvenile is similar but has duller plumage overall and a speckled crown. Utters whistling and rattling calls. Gregarious at all times; in breeding season, nests in loose colonies with all members of flock helping nesting birds. Favours woodland, including Cork Oak forests, olive groves and pine plantations. Local resident, restricted to S and central Iberia.

2. COMMON MAGPIE *Pica pica* 46cm

Distinctive long-tailed bird. Plumage is mainly black but note white belly and white on wings; latter feature is most striking in flight. In good light, note greenish-blue sheen on rounded wings and tail. Birds from NW Africa show, in addition, a patch of bare blue skin behind eye. Favours open country with scattered trees; despite frequent persecution, still occurs near human habitation. Often seen in small groups, uttering loud, rattling alarm calls. An opportunistic feeder, taking insects, fruit, animal road-kills and young birds and eggs. Widespread and common resident across S Europe and, locally, in NW Africa.

3. EURASIAN JAY *Garrulus glandarius* 34cm

Distinctive and colourful bird. Birds from European part of range have mainly pinkish-buff body plumage; note, however, the black 'moustache' and the white undertail and rump. Crown is pale with dark streaks, and the black wings have a white patch and chequerboard patch of blue, black and white. Birds from Middle East are paler with a dark cap. Birds from NW Africa are darker but also show a dark cap. Call is a loud, raucous *kraah*; song is a subdued series of harsh notes and mewing sounds. Favours broadleaved woodland but also occurs in pine forests. Buries acorns where oaks are present. Generally shy. Widespread resident across S Europe; local in NW Africa and the Middle East.

4. WESTERN JACKDAW *Corvus monedula* 33cm

Familiar small corvid. Has mainly smoky-grey plumage, darkest on cap and wings and palest on nape. Iris is pale in adult but darker in juveniles. Aerobatic in flight, frequently uttering sharp *chyack* calls. Walks with a characteristic swagger. Favours sea cliffs and farmland. Nests in tree holes, rock crevices and in buildings. Opportunistic feeder with a varied diet. Widespread resident across S Europe and, locally, in NW Africa and Middle East.

5. RED-BILLED CHOUGH *Pyrrhocorax pyrrhocorax* 40cm

Medium-sized corvid with glossy, dark plumage. Adult has bright red legs and a long, downcurved red bill; bill of juvenile is yellowish and shorter than in adult. In flight, note broad, 'fingered' wingtips and frequently uttered *chyah* call. Favours inland cliffs and rocky coasts. Probes ground for insects. Forms flocks outside breeding season. Widespread resident in Iberia; also occurs, locally, eastwards to S Turkey and in NW Africa.

6. ALPINE (YELLOW-BILLED) CHOUGH *Pyrrhocorax graculus* 37–39cm

Similar to Red-billed Chough but smaller, with proportionately longer tail, shorter legs and a yellow bill that is shorter even than juvenile Red-billed. Aerobatic, and typically seen in flocks. Often bold and inquisitive. Calls include a trilling *prrrr* and a high-pitched whistle. Locally common on mountain slopes above 1500m.

1

2

3

4

4. juv

5

6

1. COMMON RAVEN *Corvus corax* 64cm

The largest corvid in the region. Has all-dark plumage, a shaggy, ruffed throat and a massive bill; plumage has an oily sheen in good light. Often seen in flight and recognised then by its long, thick neck, the long, broad wings and the long, wedge-shaped tail. Very aerobatic, tumbling and rolling in mid-air. Utters loud, deep, *cronk* call. Favours mountains, rocky coasts, inland cliffs and rugged terrain generally. Often seen in pairs. Widespread resident across the region; least numerous in NE Africa and the Middle East.

2. FAN-TAILED RAVEN *Corvus rhipidurus* 46–50cm

Similar to Brown-necked Raven but with a shorter tail and a more massive bill. Plumage is all dark. Tail appears most noticeably short in flight, when it is often fanned out. Wings are broad, particularly towards the base, so that flight silhouette bears a passing resemblance to a miniature vulture; unlike vultures, however, Fan-tailed Ravens are extremely aerobatic. Calls include a deep *krahrr*. Favours deserts and arid mountain terrain. Local resident in S Middle East.

3. BROWN-NECKED RAVEN *Corvus ruficollis* 50–55cm

Resembles Common Raven but is smaller and less robust, and has a relatively narrow bill. Plumage is all dark but in good light note the bronze sheen on the nape. Flight silhouette is similar to that of Common Raven, with long, broad wings and a wedge-shaped tail. Calls include a harsh *kreaah*. Favours deserts and wadis. Widespread resident across N Africa and, locally, in S parts of the Middle East.

4. CARRION/HOODED CROW *Corvus corone* 45–49cm

Two races occur in the region, treated by some as separate species. Carrion Crow *C. c. corone* has all-dark plumage and occurs throughout Iberia and S France. Hooded Crow *C. c. cornix* has black head, wings and tail but otherwise grey to lilac-grey plumage; it occurs from Italy, Corsica and Sardinia eastwards to the Middle East. Call (of both races) is a coarse *coarr-coarr-coarr....* Favours farmland and open country but also visits outskirts of villages to scavenge; extremely wary of people. Generally solitary. Widespread resident throughout its range.

5. ROOK *Corvus frugilegus* 45–48cm

A rather stately member of the crow family. Adult has glossy black plumage and a bare, white facial patch at the base of the long bill. Immature birds have more matt plumage and a shorter bill that is dark at the base; hence, at this age, confusion with Carrion Crow is possible. In flight, note the relatively long wings and long tail (rounded at the tip) in all birds. Call is a nasal *carr*. Favours farmland, particularly ploughed fields and short grassland. Invariably seen in flocks. Present year-round in N of region. Otherwise typically a widespread winter visitor from N Iberia and France to Turkey (Oct–Mar).

6. HOUSE CROW *Corvus splendens* 38–40cm

A bold and distinctive member of the crow family. Plumage resembles that of Western Jackdaw but the bird is appreciably larger and has a proportionately much bigger bill. Plumage is mainly blackish except for the grey nape, neck and breast. In flight, note the relatively long tail. Call is a subdued *crarr*. Usually found around habitation and is an opportunistic scavenger. Generally seen in flocks. Resident in S Middle East and a recent arrival there from the Indian subcontinent (possibly introduced).

1

2

Hooded

3

4

4. Carrion

5

5

6

1. HOUSE SPARROW *Passer domesticus* 14–15cm

Region's most familiar sparrow. Adult male in summer has grey crown, cheeks and rump, chestnut-brown nape, back and wings, pale grey underparts and black throat. Winter male has subdued plumage colours and pattern. Female is nondescript, with streaked buff and grey-brown plumage and pale buff supercilium behind eye. (Relationship with Spanish Sparrow is complex. Male of so-called **Italian Sparrow** *P.d.italiae* – arguably a race of Spanish – resembles male House Sparrow but has pale cheeks and chestnut crown.) Calls of all sparrows include familiar sparrow chirps. Associated with human habitation. House Sparrow is a widespread resident, but absent from Italy and Sardinia, where Italian Sparrow occurs; latter also found on Corsica, Sicily, Malta and Crete.

2. DESERT SPARROW *Passer simplex* 13cm

Subtly marked and attractive sparrow. Summer male has mainly greyish-buff plumage, palest on the underparts, a black mask and throat, and black on wing coverts and primaries. In winter, black throat is less well marked and plumage is flushed overall with pinkish buff. Female (at all times) has sandy-buff plumage, palest on the underparts; wing markings are grey (not black). Calls are similar to those of House Sparrow but higher pitched. Favours desert oases. Local resident in NW Africa.

3. EURASIAN TREE SPARROW *Passer montanus* 14cm

Well-marked sparrow. Sexes are similar. Adult has a chestnut cap and nape, a small black bib and a black patch on the otherwise whitish cheeks. Upperparts are otherwise streaked chestnut-brown, but note the two white wingbars; underparts are greyish white. Juvenile has subdued plumage colours and pattern. Utters House Sparrow-like chirps but also a sharp *tik-tik* in flight. Associated with arable farming communities. Typically nests in a tree hole; occasionally may use a White Stork's nest. Widespread resident across Europe and much of Turkey.

4. DEAD SEA SPARROW *Passer moabiticus* 12cm

A tiny, well-marked sparrow. Adult male in spring has grey-brown on the head, neck and underparts, but a distinctive facial pattern comprising a black eyestripe and throat, a pale supercilium flushed yellow-brown behind the eye, and a pale stripe (flushed yellow) framing the throat and the side of the neck. Upperparts are otherwise streaked chestnut, black and buff. Male's plumage colours and pattern are subdued outside the breeding season. Female has nondescript streaked buffish-brown upperparts and buffish-white underparts. Calls are typically sparrow-like but higher pitched. Favours areas of arid scrub and cultivation in the vicinity of water. Resident and partial migrant in the Middle East.

5. SPANISH SPARROW *Passer hispaniolensis* 15cm

Adult male in summer has chestnut on the crown, nape and sides of neck, a black throat and bib, and striking white cheeks. Upperparts are otherwise boldly streaked with chestnut, black and white; shows two pale wingbars. Whitish underparts are heavily patterned with black arrowhead markings, particularly on the breast and flanks. In winter, male's plumage colour and pattern are subdued. Bill is black in breeding season but paler at other times. Female has streaked brown upperparts and faintly streaked greyish underparts; almost identical to female House Sparrow. Calls include typical sparrow-like chirps. Gregarious at all times. Nests constructed in trees, buildings and in White Stork nests. Typically a summer visitor in NE; these birds winter from the Middle East southwards and eastwards. Resident and partial migrant elsewhere. *See also* Italian Sparrow entry above.

1. *italiae*

1. ROCK SPARROW *Petronia petronia* 14cm

Resembles a female House Sparrow but has a distinctive plumage pattern. Sexes are similar. Adult has grey-brown upperparts, streaked with buff and black; note the pale crown stripe, dark stripe on the side of the crown, pale supercilium and dark eyestripe. Underparts are grey-ish white with buff streaks on breast, flanks and undertail feathers. Pale throat is defined by dark stripes; diagnostic yellow spot in centre of lower throat is often hidden. Note the pale spots at tip of tail and the stout bill with an orange base to the lower mandible. Juvenile is similar to adult but with pale fringes to the wing feathers. Calls include a nasal *vuee*; song includes call-like notes. Favours arid rocky terrain. Locally common resident in S Europe, Turkey, NW Africa and Middle East.

2. PALE ROCK SPARROW *Carpospiza brachydactyla* 15cm

A pale and extremely nondescript sparrow. Sexes and ages are similar. Upperparts are uniform pale sandy-grey (darkest on the wings) and underparts are pale buffish white. Note the indistinct pale supercilium and hint of a pale wing panel and two pale wingbars. Calls include a faint, European Bee-eater-like *pruup*; song is buzzing and insect-like. Favours arid slopes and semi-desert. Rare summer visitor and passage migrant to the Middle East, present May–Aug.

3. COMMON CHAFFINCH *Fringilla coelebs* 15cm

Familiar and colourful finch with striking white wingbars at all times and ages. Across most of the Mediterranean, adult male in summer has a reddish-pink face and underparts, a blue crown and a chestnut back; in winter, bill is pinkish (not blue) and plumage colours are subdued. Male from NW Africa (*F. c. africana*) has a greenish back, a mainly blue head, and a pinkish throat and underparts. All females are buffish brown, darker above than below, but share male's white pattern on the wings. Call is a distinct *pink-pink*; song comprises a descending trill with characteristic final flourish. Favours broadleaved and conifer woodland, farmland and gardens; forms flocks outside the breeding season. Widespread and common resident across Europe, Turkey and NW Africa; numbers increase in winter and range then extends to NE Africa and the Middle East.

4. BRAMBLING *Fringilla montifringilla* 14–15cm

Similar to Common Chaffinch but recognised at all times by the orange-buff flush to the shoulder, breast and flanks, the mainly black wings with orange-buff and white wingbars, and the white rump seen in flight. Note also the whitish belly and dark spots on the flanks. Female and immature birds have a buffish-grey face with dark, parallel lines down nape. Winter male has a dark blue-grey and brown head; this is black in breeding plumage (sometimes acquired by wintering birds before they leave the region in early spring). Calls include a harsh *eerp*. Favours open woodland (especially Beech) and often seen in flocks. Widespread winter visitor, present Nov–Mar.

5. HAWFINCH *Coccothraustes coccothraustes* 18cm

Easily recognised by its massive, conical bill. Note also the relatively large head and the short, white-tipped tail. Adult has pinkish-buff, orange-buff and chestnut elements to plumage; colours of male are brighter than those of female. Juvenile is similar to female but plumage is spotted. Large white wingbar is obvious in undulating flight in all birds. Calls include a sharp, robin-like *tik*. Favours broadleaved woodland and feeds on hard-cased seeds. Local resident in Europe and NW Africa; more widespread and numerous in winter.

africana

1. EUROPEAN GOLDFINCH *Carduelis carduelis* 12cm

Colourful and distinctive finch. Adult has red and white on face, black cap and sides of neck, a buffish back, and white underparts with buff flanks. Note also the yellow wingbars and white rump. Juvenile has brown, streaked plumage and yellow wingbars. Calls include a tinkling *tee-te-litt* or *te-litt*; twittering song includes elements of call. Favours meadows and fallow arable fields. Widespread resident (scarce in NE Africa); numbers boosted in winter by birds from N Europe.

2. SYRIAN SERIN *Serinus syriacus* 12.5cm

Adult male is overall greenish yellow but with grey wash on crown, neck and flanks. Note yellow on face, streaked back and relatively long tail. Adult female is similar to male but with more grey and less yellow on the face. Juvenile is similar to adult female but grey elements of plumage are buff. Calls include a buzzing *ter-let*; song is twittering. Favours mountain slopes in summer but orchards and semi-deserts at lower elevations in winter. Scarce summer visitor (Apr–Aug) to mountains of N Middle East; winters (Oct–Mar) further south in Middle East.

3. RED-FRONTED SERIN *Serinus pusillus* 12cm

Distinctive adult looks overall dark brown with a striking red fore-crown. On close inspection note the black head, the brown, heavily streaked upperparts, and the streaked underparts that grade from buffish brown on the breast to whitish on the vent. Also has two pale wingbars and reveals a pale buffish rump in flight. Juvenile is similar to adult but plumage is overall paler, head is brown not black, and red forecrown is absent. Calls include a rapid trill; song is fast and twittering. Favours wooded mountains in summer, but lower elevations in winter. Mainly resident from Turkey eastwards; some dispersal in winter.

4. EUROPEAN SERIN *Serinus serinus* 11.5cm

A tiny, colourful finch. Adult male has a yellow head, breast and rump. Upperparts are otherwise streaked yellowish brown, while underparts are paler and streaked. Note the two pale wingbars and forked tail. Adult female and juvenile are similar but colours are subdued (especially in juvenile). Calls include a trilling *tirililit*; song is jingling and fast. Favours orchards, open forests and gardens. Widespread resident in Europe, Turkey and NW Africa. Numbers increase in winter, when range extends to NE Africa and Middle East.

5. CITRIL FINCH *Serinus citrinella* 12cm

Tiny finch. Mainland Europe adult male is overall greenish yellow with a yellow face and grey on the neck and breast; note the two yellow wingbars and greenish-yellow rump. Sardinia and Corsica adult male is similar but back is brown and streaked. Adult females are similar to respective males but colours are duller and back is streaked. Juvenile has streaked brown plumage. Calls include a harsh *cheeht*; song is twittering and rapid. Favours forested mountains, usually close to the tree-line. Local resident in W, including Corsica and Sardinia.

6. COMMON LINNET *Carduelis cannabina* 13–14cm

Adult male has a grey head, a chestnut back and buffish-white underparts; in summer, acquires a rosy-pink flush on the forecrown and breast, while in winter, head and back appear streaked. Adult female (at all times) and juvenile resemble winter male but back is grey-brown. Calls include a sharp *tet-ett-ett*; song is twittering and musical. Favours scrub, maquis and arable fields. Widespread resident; numbers boosted in winter by migrant visitors.

1. juv

2. ♂

4. ♀

4. ♂

♂

3

6

6. juv

5

♂

♀

1. DESERT FINCH *Rhodospiza obsoleta* 13cm

Adult male has mainly sandy plumage with black lores, a black bill, and dark wings with contrasting white and pink feather margins; note also the white outer tail. Adult female and first-winter birds have paler bill and lores. Call is a rolling *trrrt*; subdued song is nasal and faltering. Favours open, arid terrain including cultivated fields. Resident in the Middle East.

2. EUROPEAN GREENFINCH *Carduelis chloris* 14–15cm

Breeding male is bright yellow-green with hint of grey on face, neck and wings; for most of year, colours are duller. Female is mainly grey-green, the back subtly streaked and washed brown. Juvenile is greenish-grey above, greyish-white below and streaked all over. All birds have a yellow wing patch and yellow rump and sides to tail. Bill is conical and pinkish in all birds. Call is a sharp *jrrrrup*; song is either a call-like wheezy *weeeish* or a whistling twitter. Favours woods and gardens. Widespread resident but mainly a winter visitor to Middle East.

3. EURASIAN SISKIN *Carduelis spinus* 12cm

Slender-billed finch with two yellow wingbars, and yellow rump and sides to tail. Summer adult male has yellow and green upperparts and pale underparts, streaked on flanks; black bib and forehead but otherwise yellow head and neck. Non-breeding male is duller and black on face is less evident. Female and first-winter birds have duller colours and face lacks black markings. Calls include a twittering *tsweee*; song is a twittering warble. Forms flocks in winter. Favours woodland. Widespread winter visitor (Oct–Mar); local resident in uplands.

4. EURASIAN BULLFINCH *Pyrrhula pyrrhula* 14–16cm

Adult male has rosy-pink face and breast, black cap and blue-grey back. Note the black wings and white wingbar, black tail and white rump. Adult female is similar to male but colours are duller. Juvenile resembles female but head and neck are uniform buffish brown. Call is a soft, piping *pyuu*; subdued warbling song includes call-like notes. Favours woodland and scrub. Local resident in N of range; more widespread in S in winter.

5. COMMON ROSEFINCH *Carpodacus erythrinus* 14–15cm

Compact finch with a thick, stubby bill. Most birds are essentially brown, streaked above and with two pale wingbars. However, adult male has red on head, breast and rump; upperparts otherwise brown and underparts whitish. Call is a whistling *tluee*; song comprises whistling phrases. Favours scrub. Occasional breeder, and scarce passage migrant in E.

6. SINAI ROSEFINCH *Carpodacus synoicus* 13–15cm

Recalls Common Rosefinch but with smaller bill and less obvious wingbars. Female and immature are greyish brown, palest on underparts. Adult male has pinkish red on face, underparts and rump. Call is a squeaky *chrrp*. Favours stony wadis. Middle East only.

7. TRUMPETER FINCH *Bucanetes githagineus* 12–13cm

Dumpy finch with a large, stubby bill. Summer adult male is buffish pink with a blue-grey head flushed red on forecrown, and bright red bill. Note the pinkish rump and upper tail, and pink margins to otherwise dark wing feathers. Winter male and female (at all times) have duller colours. Juvenile is duller still. Nasal calls and song include a strange sound like a toy trumpet. Favours stony deserts. Resident NW Africa and Middle East; local in S Iberia.

1

2

2. juv

♀

3 ♂

4 ♂

♀

5 ♂

♂

6 ♂

7

1. CROSSBILL *Loxia curvirostra* 16–17cm

Extraordinary finch with cross-tipped mandibles. Male is dull red while female is yellowish green; juvenile is streaked buffish brown. In all birds, wings are darker than other areas of plumage. Flight call is a sharp *kip-kip*; song includes call-like notes. Restricted to extensive conifer forests. Feeds high in trees but often visits pools to drink. Presence of feeding birds often indicated by sound of falling cones. Locally common resident in Europe, Turkey and NW Africa.

2. HOUSE BUNTING *Emberiza striolata* 14cm

A small bunting. Middle East adult male has a streaked, grey head and neck, the face marked with dark and whitish stripes; back is grey-brown and streaked, wings are reddish brown with dark centres to greater coverts, and underparts are buffish brown. Female and first-winter birds are similar but colours are subdued and head pattern is indistinct. NW Africa adult male has a streaked grey head with subdued paler and darker stripes on face; upperparts are otherwise uniformly reddish brown and underparts are orange-buff. Female and first-winter birds are similar but head pattern is indistinct. All birds have a dark upper mandible and pink lower one, and buffish outer tail feathers. Calls include a nasal *tchew-ee*; song is Chaffinch-like. Found around human habitation in NW Africa but on uncultivated slopes in Middle East. Resident in NW Africa and the Middle East (these two races are contenders for a species split).

3. ROCK BUNTING *Emberiza cia* 16cm

Male has a blue-grey head and breast, the face marked with black stripes. Upperparts are otherwise streaked reddish brown and underparts are reddish orange; note the two whitish wingbars. Adult female and first-winter birds are similar to male but colours and markings are subdued. Juvenile has brown, streaked plumage. All birds have a greyish bill and white outer tail feathers. Call is a sharp *tzee*; song is high-pitched, rapid and warbling. Favours warm, rocky slopes; found around the tree-line in summer but at lower elevations in winter. Mainly a local resident from Iberia to Turkey, and in NW Africa; some dispersal occurs within this overall range in winter.

4. REED BUNTING *Emberiza schoeniclus* 15cm

Well-marked bunting. Adult male in summer has a black head and throat with a striking white moustache; upperparts are chestnut, streaked with buff and black, while underparts are pale and streaked. In winter, black elements of male's plumage are less distinct, notably on head. Female and first-winter birds have streaked brown upperparts and pale streaked underparts; note also the buff supercilium, dark malar stripe and pale submoustachial stripe. All birds have white outer tail feathers. Call is a thin *tseeu*; song is a series of chinking notes, ending in a flourish. Favours marshes and farmland. Local resident but widespread winter visitor to Europe, Turkey and NW Africa.

5. YELLOWHAMMER *Emberiza citrinella* 16–17cm

Colourful adult male has a yellow head and underparts, and a chestnut back, rump and wings; the dark lines (through eye, on ear coverts and on sides of crown) are more prominent in winter than summer. Female has more subdued colours than male and darker head markings (similar to winter male). Juvenile is sandy-brown and streaked; note the chestnut rump. Calls include a nasal *tsiir*; song is a distinctive *tse-tse-tse-tse-tse-tse-tsii-tsuu*. Favours scrub-covered slopes and hedgerows in summer; in winter, often found on arable land. Locally common resident but more widespread as a winter visitor.

Picture Credits

The photographs were taken by Paul Sterry with the exception of the following (numbers are page numbers with photograph numbers in parentheses; if only the photograph number is given it refers to the main image):

Ardea: Eric Dragesco 61 (3).
FLPA: Hans Dieter Brandl 149 (5); W.S. Clark 53 (4 ad flying), 53 (4 juv flying); Silvestris Fotoservice 143 (6).
Natural Image: Mike Lane 143 (1), 143 (3).
Nature Photographers Ltd: Frank Blackburn 47 (3 female standing), 59 (1 inset standing), 59 (1), 115 (2), 119 (4), 145 (3), 183 (1 male); Mark Bolton 25 (1 flying), 25 (1 swimming), 25 (3), 45 (2 juv flying), 87 (4 2nd summer flying), 107 (4), 117 (3 flying), 129 (3), 141 (5), 179 (4 female); Derick Bonsall 125 (1); Kevin Carlson 45 (3), 47 (1 male), 67 (5), 115 (2 female), 127 (3 female *feldegg*), 141 (2), 141 (3 female), 143 (2), 153 (3), 155 (5), 155 (6), 159 (4 Cyprus), 161 (4), 183 (3 female), 183 (3), 185 (1 female), 185 (5 female); Colin Carver 23 (5 summer), 23 (7), 59 (2), 113 (3 female), 113 (3), 131 (1), 139 (1 female), 139 (1), 145 (5), 149 (1), 155 (4), 161 (2), 183 (4); Bob Chapman 99 (2); Hugh Clark 37 (2 flying), 57 (2 male), 93 (3), 105 (3 flying), 113 (3 flying), 115 (1 flying), 115 (2 flying), 117 (2 flying), 117 (4 flying), 161 (5 flying), 183 (1 female); Andrew Cleave 23 (3 winter), 25 (5 standing), 25 (5 swimming), 27 (3 swimming), 33 (4 with cow), 89 (2); Peter Craig-Cooper 59 (6), 67 (2), 93 (5); Dick Daniell 27 (4 flying); Geoff Du Feu 31 (4 flying), 53 (1 flying); Michael Gore 61 (2), 67 (5 female), 71 (3), 91 (2 ad winter), 91 (4), 111 (1 flying), 149 (6), 159 (5a), 161 (5), 167 (1); Phil Green 169 (3 juv), 18 (left background), 61 (4), 179 (6); James Hancock 69 (2 flying); Mike Hill 27 (2 drying wings), 59 (3), 125 (4), 125 (5), 127 (3 *flava*), 141 (2 female); Barry Hughes 47 (1 female standing), 53 (2), 61 (6), 75 (1 winter), 75 (4), 77 (3 winter), 81 (4), 87 (3), 89 (4), 125 (5 inset), 141 (3), 145 (1), 153 (2), 185 (1); Ernie Janes 49 (4), 51 (2), 55 (5), 63 (1 inset), 63 (1), 103 (3); Richard Mearns 81 (2), 91 (3); Lee Morgan 16 (background), 87 (4 ad flying); Philip Newman 47 (3 female flying), 47 (3 male flying), 47 (3 male), 49 (6 male), 57 (2 female flying), 57 (2 male flying), 57 (4 female), 57 (4 male), 73 (1), 77 (3 summer), 79 (2 summer), 119 (5), 125 (2), 129 (2), 133 (4 female), 133 (4), 143 (2 inset), 151 (5), 159 (1), 183 (5); David Osborn 25 (4 ad), 43 (2 female), 53 (1), 85 (5 summer), 105 (1); Bill Paton 51 (4), 103 (1); Peter Roberts 91 (5); Don Smith 103 (1 flying), 103 (2 flying); James Sutherland 14 (background); Roger Tidman 23 (4), 29 (4), 45 (1 ad flying), 45 (1 juv flying), 45 (1), 45 (3 flying), 47 (4 male flying), 47 (4 male), 49 (2), 49 (3), 49 (6 female), 51 (4 flying), 51 (5a ad), 51 (5a juv), 55 (1 juv), 55 (1), 55 (2), 55 (3), 57 (1 male flying), 61 (1), 67 (5 flying), 67 (6 female), 67 (6 flying), 67 (6), 73 (1 flying), 75 (6 juv), 79 (5), 85 (1), 89 (5 imm), 91 (1 flying), 93 (1 flying), 93 (1 winter), 93 (1), 93 (2), 99 (5), 101 (2), 101 (3 flying), 103 (2), 113 (3 juv), 113 (3 *sharpei*), 117 (1 flying), 119 (2), 119 (5 flying), 123 (3), 127 (3 *iberiae*), 131 (4 female), 137 (1 female), 145 (4), 147 (3 right inset), 147 (3), 149 (3), 151 (1), 151 (2 female), 151 (4 female), 153 (5 1st-winter), 155 (2), 155 (3), 155 (6 inset), 157 (2 female), 161 (2 female), 161 (3 inset), 165 (1 juv), 173 (1), 173 (3), 173 (5), 175 (2 female), 175 (2), 177 (1), 177 (5), 179 (1 juv), 179 (4), 181 (5 female), 181 (5), 185 (4 female); Patrick Whalley 171 (1); Wolmuth&Muller 49 (5 male flying), 49 (5 male), 51 (3 flying), 53 (2 flying), 55 (2 pale phase), 73 (4), 75 (5), 97 (5), 109 (3).
Windrush: Peter Basterfield 63 (2); Richard Brooks 79 (3), 147 (4); Kevin Carlson 153 (3 female); David Cottridge 73 (4 juv), 97 (1 female), 97 (3), 175 (4 female); Frederic Desmette 101 (5); Paul Doherty 51 (3); Göran Ekström 123 (1), 131 (2); Tom Ennis 59 (5); Gordon Langsbury 47 (2); Tim Loseby 179 (3); Mike McKavett 47 (2 female); Arthur Morris 57 (4 flying), 71 (4); John North

145 (6); Jari Peltomaki 83 (4); René Pop 135 (5); Roger Tidman 45 (4); David Tipling 53 (3 juv flying), 53(3 juv), 137 (5); Arnoud B. Van den Berg 51 (1), 93 (4), 137 (1); Steve Young 43 (1).

Individual photographers: Janus Andersen 49 (4 flying); David Cottridge 49 (5 female), 55 (1 flying), 63 (2 female), 89 (4 ad winter), 89 (4 imm), 93 (4 flying), 97 (1 male), 97 (2 female), 97 (2), 97 (3 female), 97 (4), 143 (3), 149 (2 main Balearic), 149 (4 male inset), 153 (1), 153 (2 female), 181 (1), 181 (6), 183 (2); Paul Doherty 97 (3 flying); Göran Ekström 147 (5), 149 (2); Dick Forsman 53 (4 ad), 53 (4 juv), 157 (4); Tim Loseby 107 (1a); Urban Olsson 61 (5); Richard Porter 177 (2); James Smith 97 (4 female), 103 (3 Middle East), 179 (2); Warwick Tarboton 63 (5); Steve Young 149 (5 imm).

Computer generated images: 9 (topography image), 23 (2 winter), 23 (6 winter), 25 (2 all images), 31 (2 flying), 31 (4), 65 (4 male), 75 (2 winter), 75 (5 male), 95 (2 non-br), 95 (3 winter), 95 (4 winter), 99 (1 both images), 99 (2 flying), 105 (2 flying), 107 (1 flying), 107 (2 flying), 107 (6), 107 (7), 113 (2 flying), 113 (5 male and female), 115 (1 *numidus*), 115 (4 male and female), 115 (5 male and female), 121 (3), 127 (3 *cinereocapilla* and *thunbergi*), 131 (3 Middle East race), 131 (4 Eastern race), 131 (5), 133 (3 N Africa race), 135 (2), 137 (3), 147 (2), 149 (4), 153 (4 female), 157 (1), 159 (3 N Africa race), 163 (1), 163 (3), 163 (4a), 167 (2), 167 (5), 171 (3 flying and Middle East race), 175 (1 *italiae*), 177 (3 *africana*), 179 (5).

Further Reading

Brooks, R. (1998) *Birding on the Greek Island of Lésvos*, Brookside Publishing, Fakenham.

Forsman, D. (1998) *The Raptors of Europe and the Middle East*, T. & A.D. Poyser, London.

Hollom, P.A.D., Porter, R.F., Christensen, S., & Willis, I. (1988) *Birds of the Middle East and North Africa*, T. & A.D. Poyser, London.

Shirihai, H., Christie, D., & Harris, A. (1996) *Birder's Guide to European and Middle Eastern Birds*, Macmillan, London.

Shirihai, H., Gargallo, G., & Helbig, A.J. (2001) *Sylvia Warblers*, Helm, London.

Sterry, P. (1994) *Field Guide to the Birds of Britain and Europe*, The Crowood Press, Ramsbury.

Sterry, P. (2000) *Complete Mediterranean Wildlife Photoguide*, HarperCollins Publishers, London.

Svennson, L., Mullarney, K., Zetterström, D., & Grant, P. (1999) *Collins Bird Guide*, HarperCollins Publishers, London.

Useful Contacts

Package holidays operate to Lésvos but do not start until early May, after the main spring migration has finished. For information about visiting the island at any time of the year, contact Richard Brooks at email@richard-brooks.co.uk or visit www.richard-brooks.co.uk. Skala Kalloni is the best place to stay on the island, from a birding perspective, and here the most birder-friendly of the numerous hotels is the Malemi; contact them on malemi@otenet.gr.

For further information on visiting southern Israel, contact James Smith at jamesp_smith@yahoo.com He is based at Kibbutz Lotan; contact them at bird-lotan@lotan.ardom.co.il or visit www.birdingisrael.com.

For information on visiting Romania's Danube Delta and Dobrogea regions, contact Daniel Petrescu at ibis@xnet.ro or visit www.ibis-tours.ro.

Index